An introduction to optimal estimation of dynamical systems

Monographs and textbooks on mechanics of solids and fluids

editor-in-chief: G. Æ. Oravas

Mechanics: Dynamical systems

editor: L. Meirovitch

1. E. H. Dowell
   Aeroelasticity of plates and shells

2. D. G. B. Edelen
   Lagrangian mechanics of nonconservative nonholonomic systems

3. John L. Junkins
   An introduction to optimal estimation of dynamical systems

An introduction to optimal estimation of dynamical systems

John L. Junkins
The University of Virginia

SIJTHOFF & NOORDHOFF INTERNATIONAL PUBLISHERS
ALPHEN AAN DEN RIJN

ISBN 90 286 0067 1

Printed in The Netherlands

To

Elouise, Stephen and Kathryn

Contents

Preface

This text is designed to introduce the fundamentals of estimation to engineers, scientists, and applied mathematicians. The level of the presentation should be accessible to senior undergraduates and should prove especially well-suited as a self study guide for practicing professionals. My primary motivation for writing this book is to make a significant contribution toward minimizing the painful process most newcomers must go through in digesting and applying the theory. Thus the treatment is introductory and essence-oriented rather than comprehensive. While some original material is included, the justification for this text lies not in the contribution of dramatic new theoretical results, but rather in the degree of success I believe that I have achieved in providing a source from which this material may be learned more efficiently than through study of an existing text or the rather diffuse literature.

This work is the outgrowth of the author's mid-1960's encounter with the subject while motivated by practical problems associated with space vehicle orbit determination and estimation of powered rocket trajectories. The text has evolved as lecture notes for short courses and seminars given to professionals at various private laboratories and government agencies, and during the past six years, in conjunction with engineering courses taught at the University of Virginia.

To motivate the reader's thinking, the structure of a typical estimation problem often assumes the following form:

- Given a dynamical system, a mathematical model is hypothesized based upon the experience of the investigator, consistent with whatever physical laws known to govern the system's behavior, the number and nature of available measurements, and the degree of accuracy desired. Such mathematical models almost invariably embody a number of poorly known parameters.
- Determine "best" estimates of all poorly known parameters so that the mathematical model provides an "optimal estimate" of the system's actual behavior.

Any systematic method which seeks to solve a problem of the above structure should generally be referred to as an estimation process. Depending upon the nature of the mathematical model of the system and the statistical properties of the measurement errors, the degree of difficulty associated with solution of such problems ranges from near-trivial to impossible.

In writing this text, I have kept in mind three principal objectives:

(1) Document the development of the central concepts and methods of optimal estimation theory in a manner accessible to most senior engineering students.

(2) Illustrate the application of the methods to problems having varying degrees of analytical and numerical difficulty. Where applicable, compare competitive approaches to help the reader develop a feel for the absolute and relative utility of various methods.

(3) Present prototype algorithms, giving sufficient detail and discussion to stimulate development of efficient computer programs, as well as intelligent use of the programs.

Consistent with objective (1), the major results are developed initially by the route requiring minimum reliance upon the reader's mathematical skills and apriori knowledge. In many cases, subsequent developments re-establish the same "end products" by alternate logical/mathematical processes. These developments lead to fresh insight and greater appreciation of the underlying theory, as well as (in many cases) construction of more powerful computational procedures.

The set of problems selected to accomplish objective (2) are typically idealized versions of real-world engineering problems. Several examples are given at the end of each chapter to illustrate the methods of that chapter. In chapters 3 and 6, however, several more significant example applications are documented. In the applications of chapters 3 and 6, the methods of the remaining four chapters are applied; often with two or more estimation strategies compared and two or more prototype models of the system considered.

In adopting objective (3), the author remains sensitive to the pitfalls of "cookbooks" for a subject as diverse as estimation. The problem solutions and algorithms are not put forth as optimal implementations of the various facets of the theory, nor will the methods succeed in solving every problem to which they formally apply. Nonetheless, it is felt that the example algorithms will prove useful, if accepted in the spirit that they are offered, namely as implementations which have proven successful in previous applications.

A significant amount of qualitative discussion and judgements are embodied in discussions of the various methods herein. Many of the most useful truths have not been rigorously established, but rather are insights gained by years of computational experience. Thus, I have not hesitated to include qualitative comments where such seems appropriate, usually near the beginning or end of theoretical developments, or in connection with numerical examples.

The material in this text spans the central core of the subject matter and provides a reasonable foundation for immediate application to a significant class of problems. The treatment is unfortunately not adequate in several important aspects. Conspicious by their absence here are the following important facets of the field: frequency domain techniques such as Wiener filters, suboptimal and reduced order Kalman filters, formal treatments of observability and controllability, Ito calculus, and a large number of allied topics from stochastic processes and optimal control theory. The constrained scope and depth of the present treatment are felt adequate for a first exposure to this subject matter. Several intermediate and advanced texts are available which treat the various specialized topics in excellent detail. The extensive bibliography, in addition to the references at the end of each chapter, provide the motivated student or practicing professional with the means for pursuing the topics which are most pertinent to the reader's applications.

I am indebted to numerous colleagues and students for contributions to various aspects of this work; these contributions are appreciated and the individual contributions are acknowledged in context. In particular, however, I wish to acknowledge the significant contributions of the following individuals: J. Blanton, C. Cohen, J. Jancaitis, L. Meirovitch, J. Saunders, S. Ray, and J. West. Finally, my heartfelt thanks and deepest appreciation must be expressed to my wife, Elouise, who did an extraordinarily efficient and accurate job of typing the manuscript.

1

Least Square Approximation

1.1 A Curve Fitting Example

Let us consider a rather idealized example to help set the stage for the developments of this chapter. Displayed in Figure 1.1.1 are measurements of some process y(t). The "physical significance" of the particular process y(t) is not of central importance in the current discussion, but it may prove useful for you to think of y(t) as one of the following phenomena:

* The lift y of a given helicopter rotor-blade assembly as a function of the rotor angular speed.

* The concentration y of carbon monoxide in the atmosphere at Los Angeles airport as a function of time t, over a particularly depressing five year period.

* The acceleration y of the Lunar Excursion Module as a function of time t, when liftoff occurs at t= 0.

Now envision a situation in which you are charged with the development of a mathematical model for y(t). In reading background literature dealing with the phenomena y(t), you uncover three significant reports authored by persons we refer to as engineers A, B, and C. Engineer A proposes that y(t) should be accurately modeled by the cubic polynomial

$$\text{Model 1:} \quad y_1(t) = c_1 t + c_2 t^2 + c_3 t^3 , \tag{1.1}$$

but engineer A was unable to determine numerical values for the constant coefficients c_i from theoretical considerations alone.

Engineer B also approached construction of a model of y(t) theoretically, but his rather well-written report led to a different proposed mathematical model for y(t):

$$\text{Model 2:} \quad y_2(t) = x_1 t^2 + x_2 \sin t + x_3 \sin 3t, \tag{1.2}$$

where, as in Model 1, engineer B was unable to determine estimates for the constants x_i from theoretical considerations.

Engineer C, who was unable to decide which (if either) of the two theoretical discussions was valid, decided to measure y(t). Figure 1.1.1 is a graph of his measurements. Engineer C's report states that "after a detailed error analysis of the measurement process, the measurement errors have been found to have a mean of $\leq 10^{-5}$, and a standard deviation of 0.5, where for large N

$$v_n = \tilde{y}(t_n) - y(t_n) = \text{nth measurement error}$$
$$= \text{(measured value)} - \text{(true value)} \tag{1.3}*$$

$$\mu = \frac{1}{N} \sum_{n=1}^{N} v_n = \text{error sample mean} \tag{1.4}$$

$$\sigma^2 = \frac{1}{N-1} \sum_{n=1}^{N} (v_n - \mu)^2 = \text{error sample variance}$$
$$= \text{(standard deviation)}^2 \text{."} \tag{1.5}$$

Now in your quest to establish a model which predicts the behavior of y(t), you might naturally attempt evaluation of the validity of the two proposed models by using some measure of "how well" eqns. (1.1) and (1.2) predict the measurements made by engineer C with "optimum" values for the constants c_i and x_i.

For the moment, continuing the discussion of the hypothetical problem solving situation, let us assume that you had read and digested the discussion in §1.2 on the *method of least squares*, and had employed a least squares algorithm to determine the optimum values for the coefficients in the two models. Upon carrying out the least square estimate calculation, you find that the

---

*In this text, measured values are denoted by (∿), unadorned symbols are used to represent the usually unknown true values.

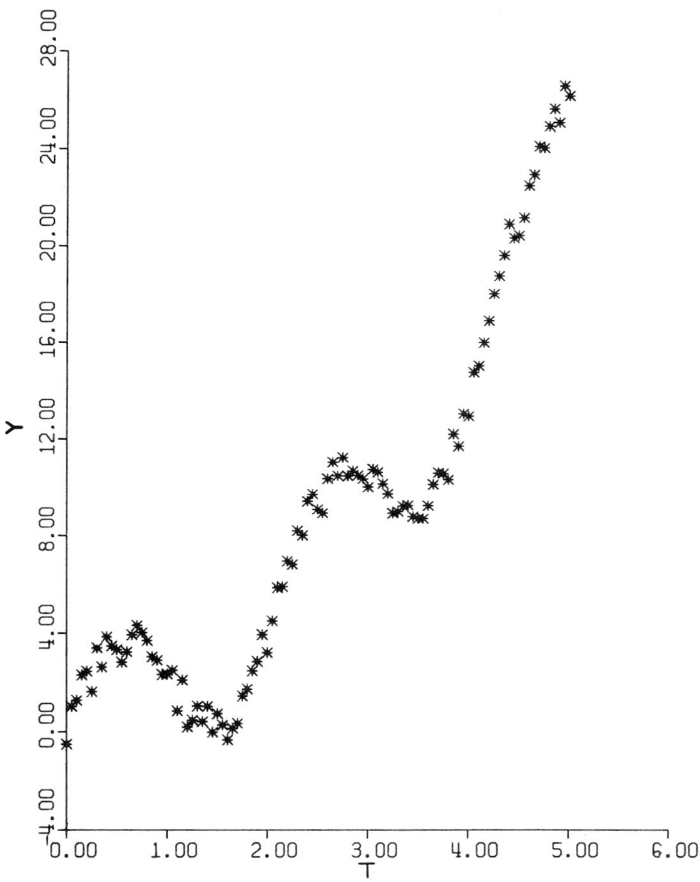

Figure 1.1.1 Measurements of $y(t)$

coefficients which bring eqn. (1.1) into best agreement with the data graphed in Figure 1.1.1 are

$$(\hat{c}_1, \hat{c}_2, \hat{c}_3) = (3.1944, -0.9774, 0.2825). \tag{1.6}$$

The solid line of Figure 1.1.2a is a graph of Model 1's best fit

$$\hat{y}_1(t) = \hat{c}_1 t + \hat{c}_2 t^2 + \hat{c}_3 t^3 \tag{1.7}$$

superimposed upon the measurements of Figure 1.1.1, and Figure 1.1.2b is a graph of Model 1's residual errors {measured data minus the best fit of eqn. (1.7)}. As is clearly evident, Model 1's best fit failed to predict significant trends in the measured data. The calculated residual error standard deviation of 2.0924 is not in good agreement with C's apriori σ of 0.5.

Carrying out the least square solution for the coefficients in eqn. (1.2) to best fit the data of Figure 1.1.1 yields the estimates

$$(\hat{x}_1, \hat{x}_2, \hat{x}_3) = (0.9944, 0.9899, 3.0465). \tag{1.8}$$

The solid line of Figure 1.1.2c is the graph of Model 2's best fit

$$\hat{y}_2(t) = \hat{x}_1 t^2 + \hat{x}_2 \sin t + \hat{x}_3 \sin 3t \tag{1.9}$$

and Figure 1.1.2d is a graph of Model 2's residual errors.

Concluding the discussion of this hypothetical example, you make the trivial qualitative observation that Model 2 is a much better representation of y(t)'s measured behavior than is Model 1. From Figure 1.1.2, you observe that Model 2's residual errors are "random" in appearance, and have a calculated mean of 10^{-6} and standard deviation of 0.5004; in excellent agreement with engineer C's apriori estimates. Having no reason to suspect that systematic errors are present in C's measurements or B's mathematical model, then your "seat of the pants" reasoning leads you to declare Model 2 "an excellent model of y(t)".

In reality, the synthetic measurements of Figure 1.1.1 were calculated by the equation

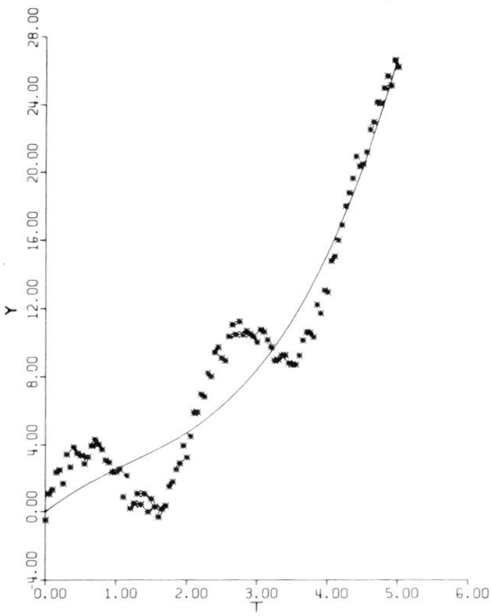

Figure 1.1.2a

Model 1 Best Fit

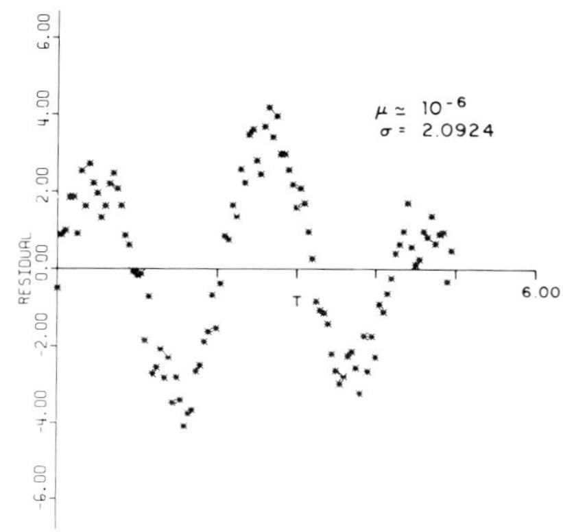

Figure 1.1.2b

Model 1 Residual Errors

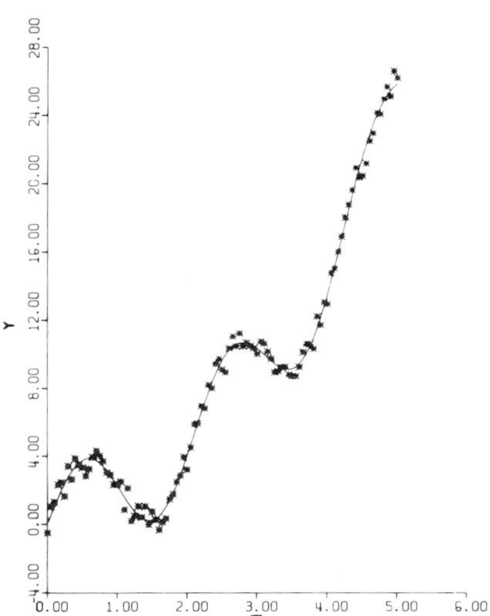

Figure 1.1.2c

Model 2 Best Fit

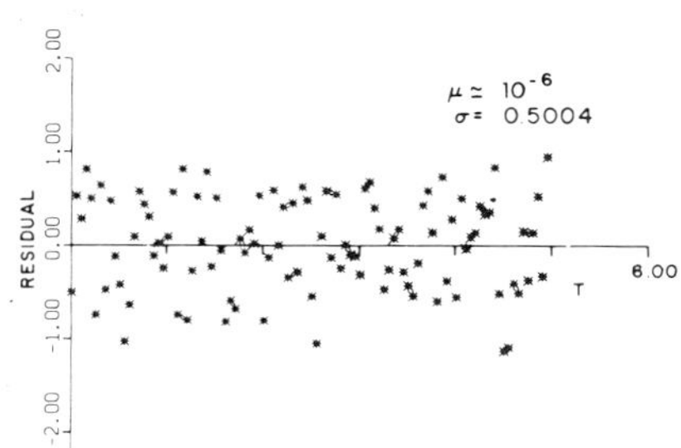

Figure 1.1.2d

Model 2 Residual Errors

$$\tilde{y}(t) = t^2 + \sin t + 3 \sin 3t + v(t) \tag{1.10}$$

where the simulated measurement errors v(t) were calculated by a normal distribution random number generator* using $\mu = 0$ and $\sigma = 0.5$. Thus it should not be surprising that Model 2 proved superior to Model 1! One important issue in estimation theory is the manner in which measurement errors propagate into errors in the estimated parameters; in the case at hand, observe that the Model 2 (which is the "true" model) optimum coefficients (1.8) differ significantly from their true values of (1, 1, 3). Also note that both models contain a t^2 term, but the corresponding numerical estimates of the t^2 coefficient are drastically different in the two best fits. In many real world problems, dominant terms in a mathematical model will have a correct mathematical structure, but higher order effects will be poorly understood. The numerical values estimated for parameters appearing in the correct portion of a model can be aliased (displaced) from their true value due to the presence of other, incorrect terms (as is clearly demonstrated above c_2's "false convergence").

In the above contrived example, Model 2 is the true equation for y(t). Generally speaking, the most challenging facet of practical applications is correctly specifying the system's mathematical model. Unfortunately, model development is the least tractable aspect of the problem setup and solution, insofar as employing universally applicable procedures. It is unlikely, indeed, that a mathematically complicated physical phenomena can be correctly modeled apriori by anyone unfamiliar with the basic principles underlying the phenomena. In short, intelligent formulation and application of estimation algorithms require intimate knowledge of the field in which the estimation problem is embedded. In numerous cases, decisions regarding which variables should be measured, the frequency with which data should be collected, the

---
*See Appendix B, and exercise 2.1

necessary measurement accuracy, and the best mathematical model can be inferred directly from theoretical analysis of the system. Estimation theory can be developed apart from considering a particular dynamical system, but successful applications almost invariably rely jointly upon understanding estimation theory and principles governing the system under consideration.

In the remaining sections of this chapter we develop fundamental aspects of estimation theory which stem from Gauss's "principle of least squares". The discussion is restricted to "static" estimation problems (the unknowns of the problem are constant parameters). We defer consideration of "dynamic" estimation problems (requiring estimation of unknown functions) until chapter 5.

1.2 Linear Batch Estimation

Bearing in mind the example discussed in §1.1, we proceed here to develop the method of least squares which will be found central to the solution of a large family of estimation problems.

Suppose that you have in hand a set (or a "batch") of measured values ($\tilde{y}_j$) of a process y(t), taken at known discrete instants of time (t_j)

$$\{\tilde{y}_1,\ t_1;\ \tilde{y}_2,\ t_2;\ \ldots;\ \tilde{y}_m,\ t_m\} \tag{1.11}$$

and a proposed mathematical model of the form

$$y(t) = x_1 a_1 + x_2 a_2 + \ldots + x_n a_n = \sum_{i=1}^{n} x_i a_i(t),\quad m \geq n \tag{1.12}$$

where

$$\{a_1(t),\ a_2(t),\ \ldots,\ a_n(t)\} \tag{1.13}$$

are a set of independent specified *basis* functions {see eqns. (1.1) and (1.2) as examples}. The x_i are a set of constants whose numerical values are unknown. It seems altogether reasonable to select the optimum x-values based upon a measure of "how well" the proposed model (1.12) predicts the measurements (1.11).

Any particular choice of x-values, say $\{\hat{x}_1, \hat{x}_2, \ldots, \hat{x}_n\}$, when substituted into (1.12) will generally not yield y-values in exact agreement with $\{\tilde{y}_1, \tilde{y}_2, \ldots, \tilde{y}_m\}$. The disagreement between observed and computed y's can stem from one or more of three possible sources:

* incorrect choice of x-values
* observation errors
* modeling errors i.e., the actual process being observed may not be accurately modeled by eqn. (1.12).

Introducing $\{e_1, e_2, \ldots, e_m\}$ as symbols for residual *errors after the solution* associated with any choice for $\{\hat{x}_i\}$, then $\{\tilde{y}_i\}$, $\{\hat{x}_i\}$ and $\{e_i\}$ must satisfy the equation

$$\tilde{y}_j = \sum_{i=1}^{n} \hat{x}_i \, a_i(t_j) + e_j, \quad j = 1, 2, \ldots, m. \tag{1.14}$$

This equation can be compactly rewritten in matrix form as

$$\tilde{Y} = A\hat{X} + E \tag{1.15}$$

where

$$\tilde{Y}^t = \{\tilde{y}_1 \; \tilde{y}_2 \; \cdots \; \tilde{y}_m\} = \text{observed (measured) y-values}$$

$$E^t = \{e_1 \; e_2 \; \ldots \; e_m\} = \text{residual errors}$$

$$\hat{X}^t = \{\hat{x}_1 \; \hat{x}_2 \; \ldots \; \hat{x}_n\} = \textit{estimated} \text{ x-values}$$

$$A = \begin{bmatrix} a_1(t_1) & a_2(t_1) & \ldots & a_n(t_1) \\ a_1(t_2) & a_2(t_2) & \ldots & a_n(t_2) \\ \vdots & \vdots & & \vdots \\ a_1(t_m) & a_2(t_m) & \ldots & a_n(t_m) \end{bmatrix}$$

and the superscript t denotes the matrix transpose operation.

Now, in the absence of model errors, the *true* y-values and *true* x-values satisfy eqn. (1.12) exactly. The *measured* y-values

$(\tilde{y}_j)$, the measurement errors (v_j, unknown), and the *true* x-values (also unknown) satisfy

$$\tilde{y}_j = \sum_{i=1}^{n} a_i(t_j)\, x_i + v_j, \quad j = 1, 2, \ldots, m \tag{1.16}$$

or, in matrix form

$$\tilde{Y} = AX + V \tag{1.17}$$

where

$$X^t = \{x_1 \ldots x_n\} = \textit{true}\text{ x-values}$$

$$V^t = \{v_1 \ldots v_n\} = \text{measurement errors.}$$

Equations (1.15) and (1.17) are identical, of course, if $\hat{X} = X$, and if the assumption of zero model errors is valid. Both of these equations are commonly referred to as the "observation equations".

Gauss's clebrated *principle of least squares** selects, as an optimum choice for the unknown parameters, the particular $\hat{x}_i$ which minimize the sum square of the residual errors

$$\phi = \sum_{j=1}^{m} e_j^2 = E^t\, E. \tag{1.18}$$

From eqn. (1.15), $E = \tilde{Y} - A\hat{X}$, and thus eqn. (1.18) can be written as

$$\phi = \tilde{Y}^t\, \tilde{Y} - \tilde{Y}^t\, A\, \hat{X} - \hat{X}^t\, A^t\, \tilde{Y} + \hat{X}^t\, A^t\, A\, \hat{X}. \tag{1.19}$$

Observe that each term of eqn. (1.19) is a scalar and because a scalar equals its transpose

$$\phi = \tilde{Y}^t\, \tilde{Y} - 2\tilde{Y}^t\, A\, \hat{X} + \hat{X}^t\, A^t\, A\, \hat{X}\ . \tag{1.20}$$

We seek to find the $\hat{X}$ which minimizes ϕ. Using the matrix calculus differentiation rules developed in Appendix A; it

*See Gauss, K.F., Theoria Motus 1809. While the principle of least squares is clearly reasonable, it's justification on probabalistic grounds will be evident in chapter 2.

follows that for a minimum we have the conditions:

necessary conditions

$$\nabla_{\hat{X}}\phi \equiv \begin{Bmatrix} \frac{\partial\phi}{\partial\hat{x}_1} \\ \vdots \\ \frac{\partial\phi}{\partial\hat{x}_n} \end{Bmatrix} = -2\ A^t\ \tilde{Y} + 2\ A^t\ A\ \hat{X} = 0 \tag{1.21}$$

sufficient condition

$$\nabla^2_{\hat{X}}\ \phi \equiv \{\frac{\partial^2\phi}{\partial x_i\ \partial x_j}\} = 2\ A^t\ A, \text{ must be positive definite*.} \tag{1.22}$$

From the necessary conditions, we have the normal equations

$$(A^t\ A)\ \hat{X} = A^t\ \tilde{Y} \tag{1.23}$$

which are often written in the inverted** form

$$\hat{X} = (A^t\ A)^{-1}A^t\ \tilde{Y} \quad . \tag{1.24}$$

Equations (1.23) and (1.24) serve as the most common basis for algorithms for solution of simple least squares problems. Eqn. (1.23) is the matrix equivalent of Gauss's original "equations of condition" which he wrote in index/summation notation. In Appendix C, a numerically superior alternative to eqn. (1.24) is developed. The explicit solution (1.24) will be seen to play a role similar to $X = A^{-1}\ Y$ in solving $Y = A\ X$ for the $m = n$ case. Generally speaking if a solution of a linear system is required, numerically superior elimination procedures are available (as preferred alternatives to computing matrix inverses).

---

*See Appendix A, for a discussion of definiteness of matrices.
**Wherever matrix inverses $()^{-1}$ are indicated in this text, an implicit assumption is that the inverse exists. Rank deficient and ill-conditioned matrices occur with sufficient frequency that the reader should expect occasional numerical difficulties in carry-out inversions by *any* algorithm. For the specific case of the least squares problem, computing the inverse in (1.24) does occasionally encounter difficulties, several numerically stable algorithms are given in Appendix C.

The least square criterion (1.18) minimized to determine $\hat{X}$ implicitly places equal emphasis upon the model (1.12) agreeing with each of the observed y-values. For the common event that the measurements are made with unequal precision, this "equal weight" approach appears logically unsound. One might intuitively select weights for each measurement which are inversely proportional to the measurements estimated precision (i.e., a measurement with zero error should be weighted infinitely, while a measurement with infinite error should be weighted zero). One satisfying choice for weights is the reciprocal of the measurement variance. Thus we are motivated to seek the estimate $\hat{X}$ of X which minimizes

$$\phi = \sum_{j=1}^{m} w_j \, e_j^2 \tag{1.25}$$

where $w_j = 1/\sigma_j^2$. More generally, one "might consider the wisdom" of finding $\hat{X}$ to minimize the quadratic function

$$\phi = \sum_{j=1}^{m} \sum_{i=1}^{m} w_{ij} \, e_i \, e_j = E^t \, W \, E, \tag{1.26}$$

which, of course, includes (1.25) as a special case, where

$W = \{w_{ij}\}$ = an mxm symmetric, positive definite matrix*.

In order that $\hat{X}$ yield a minimum of (1.26), we have the requirements:

necessary condition

$$\nabla_{\hat{X}} \, \phi = -2A^t \, W \, \tilde{Y} + 2A^t \, W \, A \, \hat{X} = 0 \tag{1.27}$$

sufficient condition

$$\nabla_{\hat{X}}^2 \, \phi = 2(A^t \, W \, A) \text{ must be positive definite.} \tag{1.28}$$

From the necessary condition, (1.27) we obtain the weighted least squares normal equations

*At this point, W is otherwise arbitrary. As will be seen in chapter 2, a justifiable optimum for W is the inverse of the measurement error convariance matrix.

$$(A^t \, W \, A) \, \hat{X} = A^t \, W \, \tilde{Y} \tag{1.29}$$

which, in principle, are invertible to obtain

$$\hat{X} = (A^t \, W \, A)^{-1} A^t \, W \, \tilde{Y} \quad . \tag{1.30}$$

While the above results have been derived in the context of a curve-fitting problem, they will be found to apply to a wide variety of linear and locally near-linear problems. Observe that the estimation problem is linear since the unknowns are contained linearly in the proposed model (1.12). Many nonlinear estimation problems can be routinely solved using successive approximation methods of §1.5.

Example 1.2

As a specific numerical example (which the student can verify with a modest amount of calculation), we use four measurements

$$\{t_j, \tilde{y}_j\} = \{1, 2.347; \; 2, 3.220; \; 3, 10.026; \; 4, 12.965\}$$

which are a subset of the 101 measurements employed in the curve fitting example of §1.1 (see Figure 1.1.1). Adopting Model 2 {eqn. (1.2)}, you can verify that the optimum least square estimates of the coefficients based upon these four measurements are

$$\begin{Bmatrix} \hat{x}_1 \\ \hat{x}_2 \\ \hat{x}_3 \end{Bmatrix} = (A^t \, A)^{-1} \, A^t \, \tilde{Y} = \begin{Bmatrix} 0.9546 \\ 0.6616 \\ 3.5554 \end{Bmatrix} \tag{1.31}$$

where, in this case, the matrices are

$$\tilde{Y}^t = \{2.347, \; 3.220, \; 10.026, \; 12.965\}$$

$$A = \begin{bmatrix} 1 & \sin 1 & \sin 3 \\ 4 & \sin 2 & \sin 6 \\ 9 & \sin 3 & \sin 9 \\ 16 & \sin 4 & \sin 12 \end{bmatrix} . \tag{1.32}$$

Substitution of the estimates (1.31) into the model (1.2) yields the optimum approximation of y(t), based upon the given four measurements. Observe the unsurprising fact that the estimates (1.31) are further from their true values (1, 1, 3) than the estimates (1.8) resulting from all 101 measurements.

A very common choice for the linearly independent basis functions (1.13) are the powers of t:

$$\{1, t, t^2, t^3, \ldots, t^n, \ldots\}$$

in which case the model (1.12) is a power series polynomial

$$y(t) = x_1 + x_2\, t + x_3\, t^2 + \ldots = \sum_{i=1}^{n} x_i\, t^{i-1} \tag{1.33}$$

The least squares coefficients estimates then follow from eqn. (1.24) with the coefficient matrix

$$A = \begin{bmatrix} 1 & t_1 & t_1^2 & \cdots & t_1^{n-1} \\ 1 & t_2 & t_2^2 & \cdots & t_2^{n-1} \\ \vdots & \vdots & \vdots & & \vdots \\ 1 & t_m & t_m^2 & \cdots & t_m^{n-1} \end{bmatrix} \tag{1.34}$$

known as the *Vandermonde matrix*.

As a second example, consider the least square estimation of the amplitudes (c_{ij}) of the harmonic series

$$\begin{aligned} y_j = c_{oo} &+ c_{11} \cos(\omega t_j) + c_{12} \sin(\omega t_j) + \ldots + \ldots \\ &+ c_{n1} \cos(n\omega t_j) + c_{n2} \sin(n\omega t_j), \\ & j = 1, \ldots, m;\ m \geq 2n + 1. \end{aligned} \tag{1.35}$$

Given $\tilde{y}_j$, t_j, $W = (W_{ij})$, and $\omega = 2\pi/T$, the desired least square estimate $(\hat{c}_{ij})$ is computable as

$$\begin{Bmatrix} \hat{c}_{oo} \\ \hat{c}_{11} \\ \hat{c}_{12} \\ \hat{c}_{21} \\ \vdots \\ \hat{c}_{n1} \\ \hat{c}_{n2} \end{Bmatrix} = (A^t\ W\ A)^{-1}\ A^t\ W\ \tilde{Y}$$

where

$$A = \begin{bmatrix} 1 & \cos(\omega t_1) & \sin(\omega t_1) \ldots \cos(n\omega t_1) & \sin(n\omega t_1) \\ 1 & \cos(\omega t_2) & \sin(\omega t_2) \ldots \cos(n\omega t_2) & \sin(n\omega t_2) \\ \vdots & \vdots & \vdots \qquad\qquad \vdots & \vdots \\ 1 & \cos(\omega t_m) & \sin(\omega t_m) \ldots \cos(n\omega t_m) & \sin(n\omega t_m) \end{bmatrix}$$

In the above case, if W is chosen as an identity matrix and the sample points $\{t_1, t_2, \ldots\}$ are symmetric about $t = T/2$, then the above matrix inverse becomes trivial because $(A^tW\ A)$ will be found to be a diagonal matrix {and equation (1.35) is clearly a truncated Fourier series}. In general, any *orthogonal* set of functions $\{a_1(t) \ldots a_n(t)\}$ satisfying

$$\delta_{ij} = \sum_{k=1}^{m} a_i(t_k)\ a_j(t_k) \qquad i, j = 1, 2, \ldots, n$$

leads to a trivial inverse of the normal equations which can be written as

$$\hat{x}_i = \left[\frac{1}{\sum_{j=1}^{m} a_i^2\ (t_j)} \right] \sum_{j=1}^{m} a_j(t_j)\ \tilde{y}_j, \quad i = 1, 2, \ldots, n. \tag{1.36}$$

On several occasions herein we will make use of orthogonal basis functions; it is not however, within the scope of this text to treat this subject comprehensively. Abramowitz and Stegun's excellent handbook summarizes a large family of orthogonal polynomials and discusses their use in approximation. Observe in eqns. (1.36), that as $m \to \infty$, the summations approach time integrals and, for the case that the $a_j(t)$ are sines and cosines, the classical integral forms for the Fourier coefficients result as a special case.

1.3 Constrained Least Square Estimation

Minimization of the weighted least square criterion function (1.26) allows relative emphasis to be placed upon the model agreeing with certain measurements more closely than others. Consider the limiting case of a perfect measurement, the corresponding (diagonal element) of the weight matrix should be ∞. Since modern computers do not permit the use of ∞ (although one can often replace ∞ by a "sufficiently large" number and get satisfactory approximations), we might be motivated to seek a rigorous means for imposing equality constraints in estimation problems.

Suppose the original observation eqn. (1.15) partitions naturally into the sub-systems $\tilde{Y}_1$ and $\tilde{Y}_2$ as

$$\begin{Bmatrix} \tilde{Y}_1 \\ \cdots \\ \tilde{Y}_2 \end{Bmatrix} = \begin{bmatrix} A_1 \\ \cdots \\ A_2 \end{bmatrix} \hat{X} + \begin{Bmatrix} E_1 \\ \cdots \\ 0 \end{Bmatrix} \tag{1.37}$$

or

$$\tilde{Y}_1 = A_1 \hat{X} + E_1 \tag{1.38}$$

and

$$\tilde{Y}_2 = A_2 \hat{X} \tag{1.39}$$

where the matrices are

$\tilde{Y}_1$ = an $m_1 \times 1$ matrix of measured y-values

A_1 = an $m_1 \times n$ known constant matrix relating Y_1 to $\hat{X}$

E_1 = an $m_1 \times 1$ matrix of residual errors associated with the particular $\hat{X}$ estimate

$\tilde{Y}_2$ = an $m_2 \times 1$ matrix of perfectly measured y-values

A_2 = an $m_2 \times n$ known constant matrix relating Y_2 to $\hat{X}$,

and the dimensions satisfy

$n \geq m_2$

$n \leq m_1$

The absence of a residual error matrix E_2 in eqn. (1.37) reflects the fact that $A_2\hat{X}$ is required to equal $\tilde{Y}_2$ exactly. Thus we can formulate the problem as a constrained minimization problem of the type discussed in Appendix A. We seek a vector $\hat{X}$ which minimizes

$$\phi = E_1^t\ W_1\ E_1 = (\tilde{Y}_1 - A_1\ \hat{X})^t\ W_1\ (\tilde{Y}_1 - A_1\ \hat{X}), \tag{1.40}$$

subject to satisfaction of the equality constraint

$$\tilde{Y}_2 - A_2\ \hat{X} = 0 \tag{1.41}$$

Using the method of Lagrange multipliers (Appendix A), it is equivalent to minimize the augmented function

$$\Phi = \tilde{Y}_1^t\ W_1\ \tilde{Y}_1 - 2\ \tilde{Y}_1^t\ W_1\ A_1\ \hat{X} + \hat{X}^t(A_1^t\ W_1\ A_1)\ \hat{X} + \Lambda^t(\tilde{Y}_2 - A_2\ \hat{X}) \tag{1.42}$$

where

$$\Lambda^t = \{\lambda_1\ \lambda_2 \ldots \lambda_{m_2}\}$$

is a matrix of Lagrange multipliers. As necessary conditions for constrained minimization of ϕ, we have the requirements:

$$\nabla_{\hat{X}}\Phi = -2A_1^t\ W_1\ \tilde{Y}_1 + 2(A_1^t\ W_1\ A_1)\ \hat{X} - A_2^t\ \Lambda = 0 \tag{1.43}$$

and

$$\nabla_{\Lambda}\ \Phi = \tilde{Y}_2 - A_2\ \hat{X} = 0, \quad \tilde{Y}_2 = A_2\ \hat{X}\ . \tag{1.44}$$

Solving (1.43) for $\hat{X}$ yields

$$\hat{X} = (A_1^t\ W_1\ A_1)^{-1}\ A_1^t\ W_1\ \tilde{Y}_1 + \frac{1}{2}\ (A_1^t\ W_1\ A_1)^{-1}\ A_2^t\ \Lambda \tag{1.45}$$

substitution of eqn. (1.45) into eqn. (1.44) allows solution for the Lagrange multipliers as

$$\Lambda = 2\ \{A_2\ (A_1^t\ W_1\ A_1)^{-1}\ A_2^t\}^{-1}\ \{\tilde{Y}_2 - A_2(A_1^t\ W_1\ A_1)^{-1}A_1^t\ W_1\ \tilde{Y}_1\}. \tag{1.46}$$

Finally substitution of eqn. (1.46) into eqn. (1.45) allows elimination of Λ, yielding an explicit solution for the equality constrained least square coefficient estimates as

$$\hat{X} = \bar{X} + K\,(\tilde{Y}_2 - A_2\,\bar{X}) \tag{1.47}$$

where

$$K = (A_1^t\,W_1\,A_1)^{-1}\,A_2^t\,\{A_2\,(A_1^t\,W_1\,A_1)^{-1}\,A_2^t\}^{-1} \tag{1.48}$$

and

$$\bar{X} = (A_1^t\,W_1\,A_1)^{-1}\,A_1^t\,W_1\,\tilde{Y}_1\,. \tag{1.49}$$

Observe that the first term of (1.47) is the least square estimate $\bar{X}$ of X in the absence of the constraint equations (1.39). The second term is an additive correction in which an optimal "gain matrix" K multiplied times the constraint residual $(\tilde{Y}_2 - A_2\,\bar{X})$ prior to the correction. This general form (1.47) is seen often in estimation theory and is therefore an important result. Due to the more complicated structure of eqns. (1.47), (1.48), and (1.49), in comparison to algorithms for solution of the weighted least squares problem, it often proves expedient to use a least squares solution with a large weight on the constraint equation. To illustrate this device, we consider a simple numerical example.

Example 1.3

Example 1.2 is modified in that unequal weights are associated with the measurements. The first three measurements are now assumed to have an equal variance of 1.0, but the fourth measurement is assumed perfect (zero variance), and is therefore considered to be a constraint. The equality constrained least squares solution follows from eqn. (1.47) as

$$\hat{X}^t = \{0.958143\;\;0.642777\;\;3.501555\} \tag{1.50}$$

with

$$\tilde{Y}_1^t = \{2.347\;\;3.220\;\;10.026\} \tag{1.51}$$

$$\tilde{Y}_2 = \{12.965\} \tag{1.52}$$

$$A_1 = \begin{bmatrix} 1 & \sin 1 & \sin 3 \\ 4 & \sin 2 & \sin 6 \\ 9 & \sin 3 & \sin 9 \end{bmatrix} \tag{1.53}$$

$$A_2 = \begin{bmatrix} 16 & \sin 4 & \sin 12 \end{bmatrix} \tag{1.54}$$

$$W_1 = \begin{bmatrix} 1 & & 0 \\ & 1 & \\ 0 & & 1 \end{bmatrix} \tag{1.55}$$

As an alternative to the above, formally constrained solution, one could consider using a weighted least squares solution with a "sufficiently large" weight in the (4,4) element of the weigh matrix. Thus we employ eqn. (1.30) to compute the estimate

$\hat{X} = (A^t\ W\ A)^{-1}\ A^t\ W\ \tilde{Y}$ with

$$\tilde{Y}^t = \{2.347\ 3.220\ 10.026\ 12.965\} \tag{1.56}$$

A = equation (1.32)

$$W = \begin{bmatrix} 1 & & & 0 \\ & 1 & & \\ & & 1 & \\ 0 & & & W_{44} \end{bmatrix}. \tag{1.57}$$

Table 1.1 summarizes the solution for $\hat{X}$ and the elements of the residual vector $E = \tilde{Y} - A\ \hat{X}$, for four values of W_{44}. The entries for $W_{44} = \infty$ were taken from the constrained least square estimates (1.50) resulting from eqns. (1.47).

Thus, the current problem (for all practical purposes), was easily solved using weighted least squares version of the normal equations.

The theoretical equivalence of an infinitely weighted observation to an equality constraint, from the viewpoint that eqns. (1.30) must contain eqns. (1.47) for this limiting case, is rather difficult to establish algebraically. It is possible, however,

Table 1.1 Solution of Problem 1.1 Using Weighted Least Squares

weight	parameter estimates			measurement residuals			constraint residual
W_{44}	$\hat{x}_1$	$\hat{x}_2$	$\hat{x}_3$	e_1	e_2	e_3	e_4
1	0.954568	0.661567	3.555381	0.334	-0.206	-0.124	0.100
10	0.957767	0.644550	3.507222	0.352	-0.217	-0.130	0.011
10^2	0.958106	0.642975	3.502125	0.353629	-0.218530	-0.130977	$\sim 10^{-3}$
10^3	0.958140	0.642796	3.501613	0.353818	-0.218647	-0.131047	$\sim 10^{-4}$
10^4	0.958143	0.642779	3.501564	0.353837	-0.218659	-0.131054	$\sim 10^{-5}$
⋮	⋮	⋮	⋮	⋮	⋮	⋮	⋮
∞	0.958143	0.642777	3.501555	0.353839	-0.218660	-0.131055	0

and is an intuitively pleasing truth. In practical applications, one can often obtain satisfactory solutions of constrained least square problems in a fashion analogous to this example.

1.4 Linear Sequential Estimation

In the developments of the first three sections, an implicit assumption is present, namely that all measurements are available for simultaneous ("batch") processing. In numerous real world applications, the measurements become available sequentially in subsets and, immediately upon receipt of a new data subset, it may be desirable to determine new estimates based upon <u>all</u> previous measurements (including the current subset). To simplify the initial discussion, consider only two subsets:

$$\tilde{Y}_1^t = \{\tilde{y}_{11}\ \tilde{y}_{12}\ldots\tilde{y}_{1m_1}\} \quad = \text{an } m_1 \times 1 \text{ vector of measurements} \tag{1.58}$$

$$\tilde{Y}_2^t = \{\tilde{y}_{21}\ \tilde{y}_{22}\ldots\tilde{y}_{2m_2}\} \quad = \text{an } m_2 \times 1 \text{ vector of measurements} \tag{1.59}$$

and the associated observation equations

$$\tilde{Y}_1 = A_1 X + V_1 \tag{1.60}$$

$$\tilde{Y}_2 = A_2 X + V_2 \tag{1.61}$$

where

A_1 = an $m_1 \times n$ known constant matrix of maximum rank $n \leq m_1$

A_2 = an $m_2 \times n$ known coefficient matrix

V_1, V_2 = vectors of measurement errors,

and

X = the $n \times 1$ vector of unknown parameters.

The least square estimate $\hat{X}_1$ of X based only upon the *first* measurement subset (1.58) follows from eqn. (1.30) as

$$\hat{X}_1 = (A_1^t W_1 A_1)^{-1} A_1^t W_1 \tilde{Y}_1 \tag{1.62}$$

and W_1 is an $m_1 \times m_1$ symmetric, positive definite weight matrix associated with measurements $\tilde{Y}_1$. It is possible to consider $\tilde{Y}_1$ and $\tilde{Y}_2$ *simultaneously* and determine an estimate $\hat{X}_2$ of X based upon *both* measurement subsets (1.58) and (1.59). Toward this end, we form the *merged* observation equations

$$\tilde{\mathbf{Y}}_2 = \mathbf{A}_2 X + \mathbf{V}_2 \tag{1.63}$$

where

$$\tilde{\mathbf{Y}}_2 = \begin{Bmatrix} \tilde{Y}_1 \\ \hline \tilde{Y}_2 \end{Bmatrix}, \quad \mathbf{A}_2 = \begin{bmatrix} A_1 \\ \hline A_2 \end{bmatrix}, \quad \mathbf{V}_2 = \begin{Bmatrix} V_1 \\ \hline V_2 \end{Bmatrix} \qquad \text{(1.64a, b, c)}$$

then the optimal least square estimate based upon the first two measurement subsets follow from eqn. (1.30) as

$$\hat{X}_2 = (\mathbf{A}_2^t \mathbf{W}_2 \mathbf{A}_2)^{-1} \mathbf{A}_2^t \mathbf{W}_2 \tilde{\mathbf{Y}}_2 \tag{1.65}$$

and the merged weight matrix is

$$\mathbf{W}_2 = \left[\begin{array}{c:c} W_1 & 0 \\ \hdashline 0 & W_2 \end{array}\right]. \tag{1.66}$$

The first k subsets of measurements can be processed to determine the estimate $\hat{X}_k$ of X as

$$\hat{X}_k = (\mathbf{A}_k^t \mathbf{W}_k \mathbf{A}_k)^{-1} \mathbf{A}_k^t \mathbf{W}_k \tilde{\mathbf{Y}}_k \tag{1.67}$$

where

$$\tilde{\mathbf{Y}}_k^t = \left[\tilde{Y}_1^t \,\vdots\, \tilde{Y}_2^t \,\vdots \cdots \vdots\, \tilde{Y}_k^t\right] \tag{1.68}$$

$$\mathbf{A}^t = \left[A_1^t \,\vdots\, A_2^t \,\vdots \cdots \vdots\, A_k^t\right] \tag{1.69}$$

and

$$\mathbf{W}_k = \begin{bmatrix} W_1 & & 0 \\ & W_2 & \\ & & \ddots \\ 0 & & W_k \end{bmatrix}. \qquad (1.70)^*$$

---

*Observe the block diagonal structure of $\mathbf{W}_k$. In Chapter 2 and Appendix B, we will see that an implicit assumption here is that measurement errors can be *correlated* only to other measurements belonging to the same subset.

It is clearly possible, in principle, to continue forming merged normal equations (upon receipt of each data subset) and solving for new optimal estimates as in eqn. (1.67) or, using other algorithms (Appendix C) for solving the *batch* least squares problem. The above route does not, on the k + 1th solution, take efficient advantage of the calculations done in processing the k previous subsets of data. The essence of the *sequential* approach to the least squares problem is simply to arrange calculations for the estimate $\hat{X}_{k+1}$ of X to make efficient use of the estimate $\hat{X}_k$ and the associated side calculations.

Observe, due to the block diagonal structure of the weight matrix, that

$$\mathbf{A}_k^t \mathbf{W}_k \mathbf{A}_k = \sum_{j=1}^{k} A_j^t W_j A_j \tag{1.71}$$

and

$$\mathbf{A}_k^t \mathbf{W}_k \tilde{\mathbf{Y}}_k = \sum_{j=1}^{k} A_j^t W_j \tilde{Y}_j \tag{1.72}$$

To abbreviate notation, we define

$$\mathbf{P}_k \equiv (\mathbf{A}_k^t \mathbf{W}_k \mathbf{A}_k)^{-1} = \left(\sum_{j=1}^{k} A_j^t W_j A_j \right)^{-1} \tag{1.73}^*$$

then the estimate $\hat{X}_k$ of X, from eqn. (1.67) can be written as

$$\hat{X}_k = \mathbf{P}_k \left(\sum_{j=1}^{k} A_j^t W_j \tilde{Y}_j \right) . \tag{1.74}$$

---

*As will be evident in chapter 2, the interpretation of $\mathbf{P}_k$ as the *covariance matrix* of $\hat{X}_k$ hinges upon several assumptions, most notably, that $\mathbf{W}_k$ is the inverse of the measurement error covariance matrix. Interpreting $\mathbf{W}_k$ and $\mathbf{P}_k$ as covariance matrices is not necessary in the developments of the present chapter.

Now, the estimate $\hat{X}_{k+1}$ of X, from analogy with eqn. (1.74) can clearly be computed as

$$\hat{X}_{k+1} = P_{k+1} \left[\sum_{j=1}^{k} A_j^t W_j \tilde{Y}_j + A_{k+1}^t W_{k+1} \tilde{Y}_{k+1} \right] \tag{1.75}$$

where, from (1.73) and (1.71)

$$P_{k+1} = (A_{k+1}^t W_{k+1} A_{k+1})^{-1} \tag{1.76}$$

or

$$P_{k+1} = [A_k^t W_k A_k + A_{k+1}^t W_{k+1} A_{k+1}]^{-1} ,$$

and, from (1.73), it is clear that

$$P_k^{-1} = A_k^t W_k A_k \tag{1.77}$$

Equation (1.76) can then be written as

$$P_{k+1} = (P_k^{-1} + A_{k+1}^t W_{k+1} A_{k+1})^{-1} \tag{1.78a}$$

or

$$P_{k+1}^{-1} = P_k^{-1} + A_{k+1}^t W_{k+1} A_{k+1} . \tag{1.78b}$$

As we shall see in subsequent developments, there are a large number of arrangements of equations (1.78).

Observe from equation (1.74) that

$$\sum_{j=1}^{k} A_j^t W_j \tilde{Y}_j = P_k^{-1} \hat{X}_k \tag{1.79}$$

so that (1.75) can be written as

$$\hat{X}_{k+1} = P_{k+1} (P_k^{-1} \hat{X}_k + A_{k+1}^t W_{k+1} \tilde{Y}_{k+1}) . \tag{1.80}$$

From (1.78b), it is clear that

$$P_k^{-1} = P_{k+1}^{-1} - A_{k+1}^t W_{k+1} A_{k+1} \tag{1.81}$$

and therefore equation (1.80) reduces to a most important result in sequential estimation theory

$$\hat{X}_{k+1} = \hat{X}_k + \mathbf{P}_{k+1} A^t_{k+1} W_{k+1} \{\tilde{Y}_{k+1} - A_{k+1} \hat{X}_k\} \tag{1.82}$$

which modifies the previous best correction $\hat{X}_k$ by an additive correction to account for the information contained in the k+1th measurement subset. Equation (1.82) is a "*Kalman update equation*" for computing the improved estimate $\hat{X}_{k+1}$ where the correction term is a *Kalman gain matrix*

$$K_{k+1} \equiv \mathbf{P}_{k+1} A^t_{k+1} W_{k+1} \tag{1.83}$$

times the residual vector (k+1th measurements minus their best linear prediction). The current equation for computing $\mathbf{P}_{k+1}$, eqn. (1.78a) offers no advantage over inverting the normal equations in their original *batch* processing form {eqns. (1.75) and (1.76)}. A number of attractive rearrangements are possible, we will now set down Kalman's classical results (ref. 1.3). Following Gura's developments (ref. 1.4), we first verify the following matrix inversion identity:

$$(I_n + CD)^{-1} = I_n - C\,(I_m + DC)^{-1}\,D \tag{1.84}$$

where

C = an arbitrary n x m matrix

D = an arbitrary m x n matrix

I_n = the n x n identity matrix

I_m = the m x m identity matrix.

The proof follows:

It is clear that

$$-CD = -CD,$$

if $(I_m + DC)$ is non-singular, then

$$-CD = -C(I_m + DC)\,(I_m + DC)^{-1}\,D \tag{1.85}$$

adding $(I_N + CD)$ to both sides of (1.85) yields

$$I_n = (I_n + CD) - C(I_m + DC)\,(I_m + DC)^{-1}\,D$$

or

$$I_n = (I_n + CD) - (C + CDC)\,(I_m + DC)^{-1}\,D$$

or

$$I_n = (I_n + CD) - (I_n + CD)\,C(I_m + DC)^{-1}\,D$$

or

$$I_n = (I_n + CD)\,\left[I_n - C\,(I_m + DC)^{-1}\,D\right]. \tag{1.86}$$

If $(I_n + CD)$ is non-singular, then pre-multiplication of both sides of (1.86) by $(I_n + CD)^{-1}$ completes the proof of equation (1.84).

The significance of the above matrix inversion identity is that (if $n > m$) the inverse of a large matrix is computable from the inverse of a smaller one. In the current quest of restructuring (1.78a) to a more computationally advantageous form, the "judicious choice" of C and D in the inversion identity (1.84) are

$$C = \mathbf{P}_k\, A_{k+1}^t\, W_{k+1} \tag{1.87}$$

$$D = A_{k+1} \tag{1.88}$$

Direct substitution of (1.87) and (1.88) into (1.84) yields

$$(I_n + P_k A_{k+1}^t W_{k+1} A_{k+1})^{-1} = I_n -$$

$$P_k A_{k+1}^t W_{k+1} (I_{m_{k+1}} + A_{k+1} P_k A_{k+1}^t W_{k+1})^{-1} A_{k+1} \quad (1.89)$$

or

$$\{P_k (P_k^{-1} + A_{k+1}^t W_{k+1} A_{k+1})\}^{-1} = I_n -$$

$$P_k A_{k+1}^t W_{k+1} \{(W_{k+1}^{-1} + A_{k+1} P_k A_{k+1}^t) W_{k+1}\}^{-1} A_{k+1} \quad (1.90)$$

or

$$(P_k^{-1} + A_{k+1}^t W_{k+1} A_{k+1})^{-1} P_k^{-1} = I_n -$$

$$P_k A_{k+1}^t W_{k+1} W_{k+1}^{-1} (W_{k+1}^{-1} + A_{k+1} P_k A_{k+1}^t)^{-1} A_{k+1} \quad (1.91)$$

Post multiplication of both sides of (1.91) by P_k yields

$$(P_k^{-1} + A_{k+1}^t W_{k+1} A_{k+1})^{-1} = P_k - \quad (1.92)$$

$$P_k A_{k+1}^t (W_{k+1}^{-1} + A_{k+1} P_k A_{k+1}^t)^{-1} A_{k+1} P_k$$

Substitution of (1.92) into (1.78a) yields the Kalman "Covariance Update Equation".

$$P_{k+1} = P_k - P_k A_{k+1}^t (W_{k+1}^{-1} + A_{k+1} P_k A_{k+1}^t)^{-1} A_{k+1} P_k . \quad (1.93)$$

Thus P_{k+1} can be obtained by "updating" P_k, the updating process requires inverting an $m_{k+1} \times m_{k+1}$ matrix.

Equations (1.93) and (1.82) constitute a sequential least squares estimation algorithm *Kalman Filter* for the linear parameter estimation problem under consideration. As will be seen in chapters 3 and 6, a large number of successive applications of the recursion (1.93) occasionally introduces arithmetic errors which can invalidate the estimates (1.82). In connection with the applications of chapter 6, alternatives to (1.93) which are numerically superior will be presented.

The "State Update Equation" (1.82) can also be rearranged in several alternate forms. One of the more common is obtained as follows: Substitute (1.93) into (1.82) to obtain

$$\hat{X}_{k+1} = \hat{X}_k + \{P_k - P_k A^t_{k+1} (W^{-1}_{k+1}$$

$$+ A_{k+1} P_k A^t_{k+1})^{-1} A_{k+1} P_k\} A^t_{k+1} W_{k+1} \{\tilde{Y}_{k+1} - A_{k+1} X_k\}$$

$$= \hat{X}_k + \{ P_k A^t_{k+1} - P_k A^t_{k+1} (W^{-1}_{k+1}$$

$$+ A_{k+1} P_k A^t_{k+1})^{-1} A_{k+1} P_k A^t_{k+1}\} W_{k+1} \{\tilde{Y}_{k+1} - A_{k+1} \hat{X}_k\}$$

$$= \hat{X}_k + P_k A^t_{k+1} \{I_{m_{k+1}} - (W^{-1}_{k+1}$$

$$+ A_{k+1} P_k A^t_{k+1})^{-1} A_{k+1} P_k A^t_{k+1}\} W_{k+1} \{\tilde{Y}_{k+1} - A_{k+1} \hat{X}_k\}$$

$$= \hat{X}_k + P_k A^t_{k+1} (W^{-1}_{k+1} + A_{k+1} P_k A^t_{k+1})^{-1} \{W^{-1}_{k+1}$$

$$+ A_{k+1} P_k A^t_{k+1} - A_{k+1} P_k A^t_{k+1}\} W_{k+1} \{\tilde{Y}_{k+1} - A_{k+1} \hat{X}_k\}$$

or

$$\hat{X}_{k+1} = \hat{X}_k + \mathbf{P}_k A^t_{k+1} (W^{-1}_{k+1} + A_{k+1} \mathbf{P}_k A^t_{k+1})^{-1} \{\tilde{Y}_{k+1} - A_{k+1} \hat{X}_k\} \tag{1.94}$$

Equations (1.82) and (1.94), together with equation (1.93) constitute a "Kalman Filter" for the linear parameter estimation problem under consideration. The process can be initiated at any step by an apriori state ($\hat{X}_1$) and covariance estimate ($\mathbf{P}_1$). If apriori estimates are not available, then the first data subset can be made "sufficiently large" ($m_1 \geq n$) to start the process with

$$\mathbf{P}_1 = (A_1^t W_1 A_1)^{-1} \tag{1.95}$$

and

$$\hat{X}_1 = \mathbf{P}_1 A_1^t W_1 \tilde{Y}_1 . \tag{1.96}$$

Equations (1.93) and either (1.82) or (1.94) can be employed sequentially to estimate $\hat{X}_k$ after each measurement subset becomes available.

If the model is in fact linear and if there is no correlation between measurement errors of different measurement subsets (so that the assumed block diagonal form of $\mathbf{W}_{k+1}$ is strictly valid), then the sequential solution for $\hat{X}_{k+1}$ (equations (1.93) and (1.82)) will agree exactly with the "batch" solution (equation (1.75)), to within arithmetic errors.

Example 1.3

As a simple numerical exercise, the student should recalculate the estimates (1.31) of example 1.2 by using the sequential algorithm of equations (1.93) and (1.82) with

$$\tilde{Y}_1^t = \{2.347 \quad 3.220 \quad 10.026\}$$

and

$$\tilde{Y}_2 = \{12.965\} \quad .$$

It will also prove instructive to resolve example 1.2 using the above sequential solution with W_2 taking on ever larger values to enforce the equality constraint

$$\tilde{Y}_2 - A_2 \hat{X} = 0 \ .$$

1.5 Nonlinear Estimation: Least Square Differential Correction

It is a fact of life that most real world estimation problems are nonlinear; the preceding developments of this chapter apply rigorously to only a small subset of problems encountered in practice. Fortunately most nonlinear estimation problems can be accurately solved by a judiciously chosen successive approximation procedure. In this section, we develop the most widely used successive approximation procedure, *least square differential correction*. This method was developed by Gauss and employed to determine planetary orbits (during the early 1800's) from telescope measurements of the "line of sight angles" to the planets.

The method to be developed here is an m x n generalization of Newton's root solving method for finding the x-values satisfying $y - F(x) = 0$. As with Newton's method, convergence of the multi-dimensional generalization is guaranteed only under rather strict requirements on the functions and their first two partial derivatives, and the closeness of the starting estimates. Let us not be concerned with convergence at this stage (although be informed, convergence difficulties do occasionally occur!). Rather let us proceed with formulating the method and look at typical applications.

Assume m observable quantities modeled as

$$y_j = F_j(x_1, x_2, \ldots, x_n); \quad j=1,2,\ldots,m; \; m \geq n \tag{1.97}$$

where the $F_j(x_1, x_2, \ldots, x_n)$ are m arbitrary independent functions of the unknown parameters $\{x_i\}$; these should be interpreted as "functions" in the general sense, as specifying "whatever process one must go through" to compute the $\{y_j\}$ given the $\{x_i\}$ (including, for example, numerical solution of differential equations). We do require that $F(x_i)$ and at least its first partial derivatives be single-valued and continuous.

Suppose that a set of observed values

$$\{\tilde{y}_1, \tilde{y}_2, \ldots, \tilde{y}_m\} \tag{1.98}$$

of the variables y_j are available; it is desired to find the particular vector of x-values

$$\hat{X}^t = \{\hat{x}_1 \; \hat{x}_2 \; \ldots \; \hat{x}_n\} \tag{1.99}$$

which minimizes the weighted sum of squares of the residuals

$$\phi = \sum_{j=1}^{m} \sum_{k=1}^{m} w_{jk} \, \Delta y_j \, \Delta y_k = \Delta Y^t \, W \Delta Y \tag{1.100}$$

where

$$\Delta y_j \equiv \tilde{y}_j - F_j(x_1, x_2, \ldots, x_n) \quad j= 1,2,\ldots,m \tag{1.101}$$

$$W = (w_{jk}) = m \times m \text{ observation weight matrix,}$$

$$\Delta Y^t = \{\Delta y_1 \; \Delta y_2 \ldots \Delta y_m\} \quad .$$

In most practical problems, ϕ cannot be directly minimized by application of ordinary calculus to equation (1.100), in the sense that explicit closed form solutions result. For that reason, attention is directed to construction of a successive approximation procedure which is designed to converge to accurate least square estimates, given approximate starting values (the determination of sufficiently close starting estimates is a problem which cannot be dealt with in general, but this problem can usually be overcome, as will be seen in applications of chapter 3).

Assume that *current* estimates

$$X_c^t = \{x_{1c}\ x_{2c}\ \dots\ x_{nc}\} \tag{1.102}$$

of the unknown x-values (1.99) are available. *Whatever the unknown objective x-values* $\{\hat{x}_i\}$ *are*, they are related to their respective current estimates $\{x_{ic}\}$ by an also unknown set of corrections $\{\Delta x_i\}$ as

$$\hat{x}_i = x_{ic} + \Delta x_i; \qquad i = 1, 2, \dots, n \tag{1.103}$$

If the $\{\Delta x_i\}$ are sufficiently small, it may be possible to solve for approximations to them and thereby obtain improved estimates of $\{x_i\}$ from (1.103).

The *current* residuals corresponding to the current x-estimates X_c are calculated from (1.101) as

$$\Delta y_{jc} = \tilde{y}_j - F_j\ (x_{1c}, x_{2c}, \dots, x_{nc}), \quad j = 1, 2, \dots, m. \tag{1.104}$$

For small changes (corrections) Δx_i, the linearly predicted residuals Δy_{jp} *after the correction* follow from Taylor series expansion of (1.101) about X_c as

$$\Delta y_{jp} = \Delta y_{jc} - \sum_{i=1}^{n} \left(\frac{\partial F_j}{\partial x_i} \bigg|_c \right) \Delta x_i \qquad j = 1, 2, \dots, m. \tag{1.105}$$

These equations are conveniently written in matrix notation as

$$\Delta Y_p = \Delta Y_c - A\ \ \Delta X \tag{1.106}$$

where

$$\Delta Y_c^t = \{\Delta y_{1c} \ \dots \ \Delta y_{mc}\} \quad = \textit{current} \text{ residuals}$$

$$\Delta Y_p^t = \{\Delta y_{1p} \ \dots \ \Delta y_{mp}\} \quad = \text{linearly } \textit{predicted} \text{ residuals, } \textit{after the correction } \Delta X$$

$$A = \left(\left. \frac{\partial F_j}{\partial x_i} \right|_c \right) \quad = \text{m x n matrix of partial derivatives, evaluated with current X-estimates } X_c. \tag{1.107}$$

Recall that the objective is to minimize the weighted sum squares ϕ given by equation (1.100). The local strategy for determining the approximate corrections ("differential corrections") ΔX_i, is to select the particular corrections which lead to the *minimum sum of squares of the linearly predicted residuals*

$$\phi_p = \sum_{j=1}^{m} \sum_{k=1}^{m} w_{ij} \ \Delta y_{jp} \ \Delta y_{kp} = \Delta Y_p^t \ W \ \Delta Y_p \ . \tag{1.108}$$

Before carrying out this minimization, it is noted (to the approximation that the linearization implicit in the prediction (1.105) is valid) that minimization of ϕ_p (1.108) is equivalent to minimization of ϕ, (1.100). If the process is convergent, the ΔX determined by minimizing (1.108) would be expected to decrease on successive iterations until (on the final iterations) the linearization is an extremely good approximation.

Substitution of equation (1.106) into (1.108) yields the quadratic in ΔX

$$\phi_p = \Delta Y_p^t \ W \ \Delta Y_p = (\Delta Y_c - A \ \Delta X)^t \ W \ (\Delta Y_c - A \ \Delta X) \tag{1.109}$$

Observe that minimization of (1.109) is completely analogous to the previously minimized quadratic form (1.26), thus any algorithm for solving the weighted least squares problem directly applies to solving for ΔX in (1.109). For example, the appropriate version of the normal equations follow as in eqns. (1.27) - (1.30) as

$$\boxed{\Delta X = (A^t\ W\ A)^{-1}\ A^t\ W\ \Delta Y_c} \quad . \tag{1.110}$$

It is also possible to apply the least square algorithm of Appendix C to solve for the ΔX in equation (1.109) which minimizes ϕ_p in eqn. (1.108).

Gauss's least square differential correction algorithm is then summarized as the following sequence of operations:

Algorithm 1.5 *Gaussian Least Square Differential Correction*

(1) Input measurements and measurement weights

$$\tilde{Y}^t = \{\tilde{y}_1\ \ldots\ \tilde{y}_m\}\ ,\quad W = \{W_{ij}\}\ .$$

(2) Input parameter starting estimates

$$X_c^t = \{x_{1c}\ \ldots\ x_{nc}\}\quad .$$

(3) Compute current values of measured functions

$$Y_c = F(X_c)\quad .$$

(4) Compute matrix of partial derivatives

$$A = \left(\left. \frac{\partial F_j}{\partial\ x_i} \right|_c \right) .$$

(5) Form measured minus computed residuals

$$\Delta Y = \tilde{Y} - Y_c$$

and their current weighted sum square

$$\phi_c = \Delta Y^t\ W\ \Delta Y,$$

go to step (8) upon convergence.

(6) Determine the correction vector ΔX which minimizes the predicted residuals sum squares

$$\phi_p = (\Delta Y - A\ \Delta X)^t\ W\ (\Delta Y - A\ \Delta X)$$

from the equation (1.110), or using the methods of Appendix C.

(7) Apply the corrections from step (6), replacing previous current parameter estimates according to

$$X_c = X_c + \Delta X$$

Return to step (3).

(8) Set $\hat{X} = X_c$ of the final (if converged) iteration.

Steps (3) through (7) of above procedure are iterated until the process converges (as evidenced by ϕ_c changing negligibly on successive iterations and/or the correction vector ΔX having negligibly small elements), or unsatisfactory convergence progress is evident (e.g., a maximum allowed number of iterations is exceeded, or ϕ_c increases on successive iterations).

The above least square differential correction process, while far from fail-safe, has been successfully applied to an extremely wide variety of nonlinear estimation problems. Note that the 1 x 1 special case of algorithm 1.5 is simply Newton's root solving method. Whenever convergence difficulties occur, they usually stem from one of the following sources: (a) The initial X-estimates are too far from the minimizing $\hat{X}$ (for the nonlinearity of the particular application), resulting in the implicit local linearity assumption being too invalid to permit convergence. (b) Numerical difficulties are encountered in solving for the corrections ΔX due to (b1) arithmetic errors corrupting the particular algorithm used to calculate the ΔX, or (b2) the A matrix having fewer than n linearly independent rows or columns (rank deficient). The difficulties (a) and (b1) can usually be overcome by a resourceful analyst, however the least square criterion does not uniquely define ΔX in the (b2) case and therefore some other criterion must be employed to select ΔX.

The following example demonstrates a simple, successful application of the least square differential correction algorithm 1.5. Other example applications are given in chapter 3.

Example 1.5

Under certain approximations (see ref. 1.5), the pitch and yaw attitude dynamics of an inertially and aerodynamically symmetric projectile can be modeled via the pair of equations

$$\theta(t) = k_1 e^{\lambda_1 t} \cos(\omega_1 t + \delta_1) + k_2 e^{\lambda_2 t} \cos(\omega_2 t + \delta_2) + k_3 e^{\lambda_3 t} \cos(\omega_3 t + \delta_3) + k_4 \tag{1.111a}$$

$$\psi(t) = k_1 e^{\lambda_1 t} \sin(\omega_1 t + \delta_1) + k_2 e^{\lambda_2 t} \sin(\omega_2 t + \delta_2) + k_3 e^{\lambda_3 t} \sin(\omega_3 t + \delta_3) + k_5 \tag{1.111b}$$

where k_1, k_2, k_3, k_4, k_5, λ_1, λ_2, λ_3, ω_1, ω_2, ω_3, δ_1, δ_2, δ_3 are 14 constants which can be related to the aerodynamic and mass characteristics of the projectile and to the motion initial conditions. These constants are often estimated by least square differential correction to "best fit" measured pitch and yaw histories modeled by eqns. (1.111).

As an example of such a data reduction process, consider the simulated measurements of $\theta(t)$ and $\psi(t)$ given in Table 1.5.1 and the *a priori* constant estimates in Table 1.5.2.

Using algorithm 1.5 (above), the inverted normal equations

$$\Delta X = (A^t WA)^{-1} A^t W\Delta Y \tag{1.112}$$

solved iteratively for corrections to the 14 parameters (using the starting estimates given in Table 1.5.2) results in the convergence history summarized in Table 1.5.3.

TABLE 1.5.1

SIMULATED OBSERVATIONS OF θ(t) AND ψ(t)

Time (sec)	$\tilde{\theta}$(Radians)	$\tilde{\psi}$(Radians)
1.	.285335E+00	.131323E+00
2.	.172814E+00	.198236E+00
3.	.686539E-01	.193740E+00
4.	.898541E-02	.152827E+00
5.	-.112413E-01	.119750E+00
6.	-.239942E-01	.108061E+00
7.	-.520285E-01	.100353E+00
8.	-.871015E-01	.716618E-01
9.	-.100698E+00	.174498E-01
10.	-.739797E-01	-.349644E-01
11.	-.202802E-01	-.529475E-01
12.	.244686E-01	-.265816E-01
13.	.301535E-01	.205463E-01
14.	-.354301E-02	.502292E-01
15.	-.464006E-01	.412986E-01
16.	-.647544E-01	.432659E-02
17.	-.506016E-01	-.298255E-01
18.	-.228399E-01	-.394071E-01
19.	-.693597E-02	-.280199E-01
20.	-.102064E-01	-.180775E-01
21.	-.200939E-01	-.273449E-01
22.	-.126085E-01	-.489978E-01
23.	.178504E-01	-.614557E-01
24.	.543561E-01	-.456612E-01
25.	.718609E-01	-.567689E-02

Variance of θ = variance of ψ measurements = $(0.0002 \text{ radians})^2$

For the problem at hand, the necessary matrices in (1.112) are defined as

$$\underset{(1\times 14)}{X^t} = \{k_1\ k_2\ k_3\ k_4\ k_5\ \lambda_1\ \lambda_2\ \lambda_3\ \omega_1\ \omega_2\ \omega_3\ \delta_1\ \delta_2\ \delta_3\} \tag{1.113}$$

$$\underset{(1\times 50)}{\tilde{Y}^t} = \{\tilde{\theta}(1)\ \tilde{\psi}(1)\ \tilde{\theta}(2)\ \tilde{\psi}(2)\ \ldots\ \tilde{\theta}(25)\ \tilde{\psi}(25)\} \tag{1.114}$$

$$\underset{(50\times 14)}{A} = \begin{bmatrix} \left.\dfrac{\partial\theta(1)}{\partial x_1}\right|_C & \cdots & \left.\dfrac{\partial\theta(1)}{\partial x_{14}}\right|_C \\ \left.\dfrac{\partial\psi(1)}{\partial x_1}\right|_C & \cdots & \left.\dfrac{\partial\psi(1)}{\partial x_{14}}\right|_C \\ \vdots & & \vdots \\ \left.\dfrac{\partial\theta(25)}{\partial x_1}\right|_C & \cdots & \left.\dfrac{\partial\theta(25)}{\partial x_{14}}\right|_C \\ \left.\dfrac{\partial\psi(25)}{\partial x_1}\right|_C & \cdots & \left.\dfrac{\partial\psi(25)}{\partial x_{14}}\right|_C \end{bmatrix} \tag{1.115}$$

$$\underset{(50\times 50)}{W} = 10^8 \begin{bmatrix} .25 & & & 0 \\ & .25 & & \\ & & \ddots & \\ 0 & & & .25 \end{bmatrix} \tag{1.116}$$

and the required 28 partial derivative expressions {needed to fill the A-matrix (1.115)} are derived directly from eqns. (1.111) as

TABLE 1.5.2

STARTING AND TRUE PARAMETER VALUES FOR EXAMPLE 1.5.1

Constant	Starting Value	True Value
k_1 =	0.5	0.2
k_2 =	0.25	0.1
k_3 =	0.125	0.05
k_4 =	0.0	0.0001
k_5 =	0.0	0.0001
λ_1 =	-0.15	-0.1
λ_2 =	-0.06	-0.05
λ_3 =	-0.03	-0.025
ω_1 =	0.26	0.25
ω_2 =	0.55	0.5
ω_3 =	0.95	1.0
δ_1 =	0.01	0.0
δ_2 =	0.01	0.0
δ_3 =	0.01	0.0

TABLE 1.5.3

CONVERGENCE HISTORY FOR LEAST SQUARE DIFFERENTIAL CORRECTIONS

Parameter	Iteration Number				
	0	1	2	...	5
k_1	0.50	0.1700	0.1949		0.1995
k_2	0.25	0.1144	0.1021		0.1002
k_3	0.125	0.0596	0.0508		0.0501
k_4	0.0	0.0000	0.0003		0.0002
k_5	0.0	-0.0017	-0.0004		0.0002
λ_1	-0.15	-0.1204	-0.0939		-0.0998
λ_2	-0.06	-0.0683	-0.0601		-0.0501
λ_3	-0.03	-0.0416	-0.0349		-0.0253
ω_1	0.26	0.2469	0.2456		0.2500
ω_2	0.55	0.5351	0.4983		0.5001
ω_3	0.95	0.9695	1.0070		1.0000
δ_1	0.01	0.0401	0.0268		-0.0002
δ_2	0.01	-0.0787	-0.0224		-0.0016
δ_3	0.01	0.0023	-0.0693		0.0002
Weighted Sum Square of Residuals					
ϕ	1.47×10^7	5.8×10^5	2.7×10^4		4×10^1

$$\frac{\partial \theta(t_j)}{\partial k_i} = e^{\lambda_i t_j} \cos(\omega_i t_j + \delta_i)$$

$$i = 1, 2, 3$$

$$\frac{\partial \psi(t_j)}{\partial k_i} = e^{\lambda_i t_j} \sin(\omega_i t_j + \delta_i)$$

$$\frac{\partial \theta(t_j)}{\partial k_4} = 1$$

$$\frac{\partial \psi(t_j)}{\partial k_4} = 0$$

$$\frac{\partial \theta(t_j)}{\partial k_5} = 0$$

$$\frac{\partial \psi(t_j)}{\partial k_5} = 1$$

$$\frac{\partial \theta(t_j)}{\partial \lambda_i} = t_j\, k_i\, e^{\lambda_i t_j} \cos(\omega_i t_j + \delta_i)$$

$$i = 1, 2, 3$$

$$\frac{\partial \psi(t_j)}{\partial \lambda_i} = t_j\, k_i\, e^{\lambda_i t_j} \sin(\omega_i t_j + \delta_i)$$

$$\frac{\partial \theta(t_j)}{\partial \omega_i} = -\, t_j\, k_i\, e^{\lambda_i t_j} \sin(\omega_i t_j + \delta_i)$$

$$i = 1, 2, 3$$

$$\frac{\partial \psi(t_j)}{\partial \omega_i} = t_j\, k_i\, e^{\lambda_i t_j} \cos(\omega_i t_j + \delta_i)$$

$$\frac{\partial\theta(t_j)}{\partial\delta_i} = - k_i \, e^{\lambda_i t_j} \sin (\omega_i \, t_j + \delta_i)$$

$$i = 1, 2, 3$$

$$\frac{\partial\psi(t_j)}{\partial\delta_i} = k_i \, e^{\lambda_i t_j} \cos (\omega_i \, t_j + \delta_i) \, .$$

Since the weight matrix (1.116) is diagonal and of the form αI, a short cut can be taken in computing the differential corrections (1.112). Observe that

$$W = \alpha \begin{pmatrix} 1 & & 0 \\ & 1 & \\ & & \cdot \\ & & & \cdot \\ 0 & & & 1 \end{pmatrix} = \alpha I$$

then eqn. (1.112) becomes

$$\Delta X = \{A^t \, (\alpha I) \, A\}^{-1} \, A^t \, (\alpha I) \, \Delta Y$$

or

$$\Delta X = (A^t \, A)^{-1} \, A^t \, \Delta Y \qquad (1.117)$$

Thus, the *unweighted* version of the normal equations can be employed to compute corrections (which seems reasonable, in the present situation in which all observations are equally weighted). We shall see in §2.4 that two interesting classes of weight matrices yield identical solutions of the normal equations, but the above case is the most frequently occurring example.

Observe the rather dramatic convergence progress evident in Table 1.5.3 (as evidenced by the decrease of the residuals weighted sum of squares by six orders of magnitude in five iterations). Also observe that the final converged values of the fifth iteration are in reasonable agreement with their respective true values listed in Table 1.5.2.

1.6 Remarks

With some reluctance, the curve fitting example of § 1.1 was presented prior to discussion of the methods of § 1.2 necessary to carry out the calculations. On several subsequent occasions herein, theoretical development of *methods* follow typical *results*, to provide motivation and to allow some apriori evaluation by the reader of the role played by the methodology under development.

The results developed in § 1.2 are among the most important in estimation theory. Indeed, the bulk of estimation theory could be viewed as extensions, modifications, and generalizations of these basic results, to address a wider variety of mathematical models and measurement strategies. We shall see, however, that the results of § 1.2 can be placed upon a more rigorous foundation and several important new insights gained through study of the developments of chapter 2 and appendixes B and C.

The sequential estimation results in § 1.4 are the simplest version of a class of procedures known as *Kalman Filter* algorithms. This methodology has been developed during the past fifteen years; the most fundamental results being published originally by Kalman (ref. 1.3) and extended/applied by numerous other investigators. A substantial portion of the present text deals with sequential estimation methodology and applications thereof.

The constrained least squares results of § 1.3 were developed originally by Junkins (ref. 1.6), and applied to post-flight estimation of rocket trajectories. The constrained least squares solution (eqn. (1.47)) is closely related to the sequential estimation solution (e.g., eqn. (1.94)), and can in fact be obtained from it by limiting arguments (allowing the weight of the constraint "observation" equations to approach infinity).

The differential correction procedures documented in §1.5 are most fundamental whenever estimation methods must be applied to a nonlinear problem. It is interesting to note that the original estimation problem motivating Gauss (i.e., determination of the planetary orbits from telescope/sextant observations) was nonlinear, and his methods (essentially §1.5) have survived as a standard operating procedure to this day. Other *mathematical programming* (Appendix A) methods are occasionally employed in minimizing the sum square residuals.

Based upon this author's experience, Gauss's algorithm should be tried first, and resort to other minimization procedures only if convergence difficulties are encountered.

1.7 *Exercises*

1.1 Following the notation of § 1.2 consider the m scalar observation equations

$$\tilde{y}_j = x + v_j \qquad j = 1, 2, \ldots m$$

or

$$\tilde{y}_j = \hat{x} + e_j \qquad j = 1, 2, \ldots m$$

which can be written in matrix form as

$$\tilde{Y} = Ax + V$$

or

$$\tilde{Y} = A\hat{x} + E$$

with $A^t = \{1\ 1\ \ldots\ 1\}$.

These observation equations hold for the simplest situation in which an unknown parameter x is *directly* measured m times (assume the measurement errors have zero means and known, equal variances). From the normal equations (1.24), establish the well known truth that the optimum least square estimate $\hat{x}$ of x is the sample mean

$$\hat{x} = (\sum_{i=1}^{m} \tilde{y}_i)/m.$$

1.2 Using the simple model

$$y = x_1 + x_2 \sin 10t + x_3\ e^{2t^2}$$

with $x_1 = x_2 = x_3 = 1.0$,

generate four sets of "synthetic data" at the instants

$$t = 0,\ 0.1,\ 0.2,\ 0.3,\ \ldots\ 1.0$$

by truncating each y value after 6, 4, 2, and 1 significant figures, respectively, to simulate (crudely) measurement errors.

Use the normal equations (1.24) to process the measurements and derive $\hat{x}_i$ estimates for each of the four cases. Compare the estimates with the true values (1, 1, 1) in each case.

1.3 Use the sequential estimation algorithm (1.93) and (1.94) to process the first three measurements of exercise (1.2) as a single measurement subset and then consider the remaining measurements to become available one at a time, for each of the four synthetic data sets of exercise 1.2.

1.8 References for Chapter 1

1.1 Gauss, K.F., Theory of the Motion of the Heavenly Bodies Moving About the Sun in Conic Sections, Dover Publications, Inc., reprinted 1963.

1.2 Abramowitz, M. and L. Stegun, Handbook of Mathematical Functions, U.S. Dept. of Commerce, NBS Applied Mathematical Series, No. 55, 9th ed., 1970.

1.3 Kalman, R.E. and R.S. Bucy, "New Results in Linear Filtering and Prediction Theory", J. Basic Engr., Trans. ASME, Vol. 83D, pp 95-108, 1961.

1.4 Gura, I.A., "An Algebraic Approach to Optimal Estimation", Hughes Space Systems Division Report No. SSD 70072R, El Segundo, California, 1967.

1.5 Nicolaides, J.D., Lectures on Free Flight Missile Dynamics, Dept. of Aerospace Engineering, Univ. of Notre Dame, 1967.

1.6 Junkins, J.L., "On the Optimization and Estimation of Powered Rocket Trajectories Using Parametric Differential Correction Processes", McDonnell Douglas Astronautics Co., Santa Monica, Calif., Report No. SM G1793, 1969.

2

Minimal Variance Estimation

2.1 *Preliminary Remarks*

The intuitively reasonable *principle of least squares* was put forth in §1.1 and employed as the starting point for all developments of chapter 1. In the present chapter, several alternative paths are followed to essentially the same basic mathematical conclusions as chapter 1. The primary function of the present chapter is to place the results of chapter 1 upon a more rigorous (or at least, a better understood) foundation. However, a number of new and computationally most useful extensions of the estimation results of chapter 1 result from the developments. The elusive weight matrix will be rigorously identified as the observation covariance matrix and some most important *nonuniqueness properties* developed in §2.5. Methods for rigorously accounting for apriori parameter estimates and their uncertainty will also be developed.

Familarity with basic concepts in probability theory is necessary for comprehension of the material in the present chapter. Should the reader anticipate or encounter difficulty in following the developments, Appendix B provides an adequate review of the concepts needed herein.

2.2 *Minimal Variance Estimation (without apriori State Estimates)*

Assume a linear model

$$\underset{(m \times 1)}{\tilde{Y}} = \underset{(m \times n)}{A} \; \underset{(n \times 1)}{X} + \underset{(m \times 1)}{V} \tag{2.1}$$

It is desired to estimate X as a linear combination of the observations $\tilde{Y}$ as

$$\underset{(n \times 1)}{\hat{X}} = \underset{(n \times m)}{M} \; \underset{(m \times 1)}{\tilde{Y}} + \underset{(n \times 1)}{N} . \tag{2.2}$$

An "optimum" choice of the matrices M and N is desired. The minimum variance definition of "optimum" M and N matrices is that the variances of *all n* estimates ($\hat{x}_i$) from their respective "true" values (x_i)

$$\phi_i = E\{(\hat{x}_i - x_i)^2\} \quad i = 1,2, \quad ,n^* \tag{2.3}$$

be minimized. This clearly requires n minimizations depending upon the same M and N; it may not be clear at this point that the problem is well-defined and whether or not M and N exist (or can be found if they do exist) to accomplish these n minimizations.

If the linear model (2.1) is strictly valid, then, for the special case of perfect measurements (V = 0) then the model (2.1) should be exactly satisfied by the perfect measurements (Y) and the true state (X) as

$$\tilde{Y} \equiv Y = AX \tag{2.4}$$

---

*E{ } denotes *"expected value"* of { }, see Appendix B.

An obvious requirement upon the desired estimator (2.2) is that perfect measurements should result (if a solution is possible) in $\hat{X} = X =$ true state. Thus, this requirement can be written substituting $\hat{X} = X$ and $\tilde{Y} = AX$ in (2.2), as

$$X = MAX + N \tag{2.5}$$

From which we conclude that M and N satisfy the constraints

$$N = 0 \tag{2.6}$$

and

$$\begin{aligned} MA &= I \\ A^t M^t &= I. \end{aligned} \tag{2.7}$$

Equation (2.6) is certainly useful information! The desired estimator then has the form

$$\hat{X} = M\tilde{Y}. \tag{2.8}$$

We are now concerned with determining the optimum choice of M which accomplishes the n minimizations of (2.3), subject to the constraint (2.7).

Subsequent manipulations will be greatly facilitated by partitioning the various matrices as follows:

The unknown M-matrix is partitioned by rows as

$$M = \begin{Bmatrix} M_1 \\ M_2 \\ \vdots \\ M_n \end{Bmatrix}, \qquad M_i \equiv \{M_{i1}\ M_{i2}\ \ldots\ M_{im}\} \tag{2.9}$$

or

$$M^t = \{M_1^t\ M_2^t\ \ldots\ M_n^t\}.$$

The identity matrix can be partitioned by rows and columns as

$$I = \begin{Bmatrix} I_1^r \\ I_2^r \\ \vdots \\ I_n^r \end{Bmatrix} = \left\{ I_1^c \; I_2^c \; \dots \; I_n^c \right\} ; \quad \text{note } I_i^r = (I_i^c)^t \; . \tag{2.10}$$

The constraint eqn. (2.7) can now be written as

$$A^t \; M_i^t = I_i^c \qquad ; \; i = 1, 2, \dots, n \tag{2.11a}$$

or

$$M_i \; A = I_i^r \qquad ; \; i = 1, 2, \dots, n \tag{2.11b}$$

and the ith element of $\hat{X}$ from (2.8) can be written as

$$\hat{x}_i = M_i \; \tilde{Y} \quad ; \qquad i = 1, 2, \dots, n. \tag{2.12}$$

A glance at (2.12) reveals that x_i depends *only* upon the elements of M contained in the ith row. A similar statement holds for the constraint equations (2.11), the elements of the ith row are independently constrained. This "uncoupled" nature of (2.11) and (2.12) is the key feature which allows one to carry out the n "separate" minimizations of (2.3).

The ith variance (2.3) to be minimized, upon substituting (2.12), can be written as

$$\phi_i = \{E \; (M_i \; \tilde{Y} \; - x_i)^2\} \; ;$$

$$i = 1, 2, \dots, n \tag{2.13}$$

Substitution of the observation eqns. (2.1) into (2.13) yields

$$\phi_i = E \; \{(M_i \; A \; X + M_i \; V - x_i)^2\} \; ;$$

$$i = 1, 2, \dots, n \tag{2.14}$$

Incorporating the constraint from (2.11) into (2.14) yields

$$\phi_i = E((I_i^r X + M_i V - x_i)^2); \quad i = 1, 2, \ldots, n \tag{2.15}$$

But $I_i^r X = x_i$, so that (2.15) reduces to

$$\phi_i = (E(M_i V)^2) \quad i = 1, 2, \ldots, n \tag{2.16}$$

which can be rewritten as

$$\phi_i = E\ (M_i (VV^t)\ M_i^t) \quad i = 1, 2, \ldots, n \tag{2.17}$$

but the only random variable on the RHS of eqn. (2.17) is V; introducing the covariance matrix of measurement errors

$$\Lambda_{vv} = E\ (VV^t), \tag{2.18}$$

then (2.17) reduces to

$$\phi_i = M_i\ \Lambda_{vv}\ M_i^t\ ; \quad i = 1, 2, \ldots, n. \tag{2.19}$$

The ith constrained minimization problem can now be stated as: Minimize each of eqns. (2.19) subject to the corresponding constraint eqn. (2.11). Using the Method of LaGrange multipliers (Appendix A), the ith augmented function is introduced as

$$\Phi_i = M_i \Lambda_{vv} M_i^t + \lambda_i^t\ (I_i^c - A^t M_i^t)\ , \quad i = 1, 2, \ldots, n \tag{2.20}$$

where

$$\lambda_i^t = (\lambda_{1_i} \lambda_{2_i} \ldots \lambda_{n_i}); \quad i = 1, 2, \ldots n \tag{2.21}$$

are n vectors of Lagrange multipliers.

The necessary conditions for (2.20) to be minimized are then

$$\nabla_{M_i^t} \Phi_i = 2\Lambda_{vv} M_i^t - A\lambda_i = 0; \quad i = 1, 2, \ldots, n \tag{2.22}$$

$$I_i^c - A^t M_i^t = 0, \text{ or } M_i A = I_i^r; \quad i = 1, 2, \ldots, n \tag{2.23}$$

From (2.22), we obtain

$$M_i = \tfrac{1}{2}\lambda_i^t A^t \Lambda_{vv}^{-1}\ ; \quad i = 1, 2, \ldots, n \tag{2.24}$$

Substitution of (2.24) into the second of (2.23) yields

$$\lambda_i^t = 2I_i^r(A^t \Lambda_{vv}^{-1} A)^{-1} \tag{2.25}$$

Therefore, substituting (2.25) into (2.24), the n rows of M are given by

$$M_i = I_i^r(A^t \Lambda_{vv}^{-1} A)^{-1} A^t \Lambda_{vv}^{-1} \; ; \; i = 1,2,..,n \tag{2.26}$$

It then follows that

$$M = (A^t \Lambda_{vv}^{-1} A)^{-1} A^t \Lambda_{vv}^{-1} \tag{2.27}$$

and the desired estimator (2.8) then has the final form

$$\hat{X} = (A^t \Lambda_{vv}^{-1} A)^{-1} A^t \Lambda_{vv}^{-1} \tilde{Y} \tag{2.28}$$

which is referred to as the Gauss-Markov Theorem.

The minimal variance estimator (2.28) is identical to the least squares estimator (1.30), provided that the weight matrix is identified as the inverse of the observation error covariance matrix.

The "sequential least square estimation" results of §1.4 are seen to embody as special case "sequential minimal variance estimation"; it is simply necessary to employ Λ_{vv}^{-1} as W in the sequential least square formulation.

2.3 Minimal Variance Estimation (with apriori state information)

The preceding results will now be extended to allow rigorous incorporation of apriori estimates ($\hat{X}_a$) of the state and associated apriori error covariance matrix (Λ_{xx_a}).

Assume the linear observation model

observed		true	observation errors	observation error covariance matrix
(mx1)	(mxn)	(nx1)	(mx1)	
$\tilde{Y}$ =	A	X +	V_y	; $\Lambda_{vv} = E(V_y V_y^t)$, known

(2.29)

and

$$\underset{\substack{\text{apriori}\\ \text{state}\\ \text{estimates}}}{\overset{(n \times 1)}{\hat{X}_a}} = \underset{\substack{\text{true}\\ \text{state}}}{\overset{(n \times 1)}{X}} + \underset{\substack{\text{errors in}\\ \text{apriori}\\ \text{estimates}}}{\overset{(n \times 1)}{V_x}} \quad ; \quad \underset{\substack{\text{apriori state's}\\ \text{error covariance}\\ \text{matrix}}}{\Lambda_{xx_a} = E(V_x V_x^t)}, \text{ known} \tag{2.30}$$

It is assumed that $\Lambda_{xv} = \Lambda_{vx}^t = E(V_y\ V_x^t) = 0$.

It is desired to estimate X as a linear transformation of the measurements $\tilde{Y}$ and the apriori state estimates $\hat{X}_a$ as

$$\hat{X} = M\ \tilde{Y} + N\ \hat{X}_a + R \tag{2.31}$$

An "optimum" choice of the $\overset{(nxm)}{M}$, $\overset{(nxn)}{N}$, and $\overset{(nx1)}{R}$ matrices is desired. As before, we adopt the minimal variance definition of "optimum" to determine the M, N, and R for which the variances of all n estimates $(\hat{x}_i)$ from their respective true values (x_i)

$$\phi_i = E\ (\hat{x}_i - x_i)^2; \quad i = 1, 2, \ldots, n \tag{2.32}$$

are minimized.

If the linear model (2.29) is strictly valid, then for the special case of perfect measurements $(V_y = 0)$, then the perfect measurements (Y) and the true state (X) should satisfy (2.29) exactly as

$$Y = AX \tag{2.33}$$

If, in addition, the apriori state estimates are also perfect $(\hat{X}_a = X,\ V_x = 0)$; an obvious requirement upon the estimator eqn. (2.31) is that it yields the true state as

$$X = MAX + NX + R \tag{2.34}$$

or

$$X = (MA + N)\ X + R. \tag{2.35}$$

Equation (2.35) indicate that M, N, and R must satisfy the constraints

$$R = 0 \tag{2.36}$$

and

$$MA + N = I \text{ or } A^t\ M^t + N^t = I. \tag{2.37}$$

Because of (2.36), then the desired estimator (2.31) has the form

$$\hat{X} = M\tilde{Y} + N\hat{X}_a \tag{2.38}$$

It is useful in subsequent developments to partition M, N, and I as follows:

$$M = \begin{Bmatrix} M_1 \\ M_2 \\ \vdots \\ M_n \end{Bmatrix},\ M^t = \{M_1^t\ M_2^t\ \ldots\ M_n^t\} \tag{2.39}$$

$$N = \begin{Bmatrix} N_1 \\ N_2 \\ \vdots \\ N_n \end{Bmatrix},\ N^t = \{N_1^t\ N_2^t\ \ldots\ N_n^t\} \tag{2.40}$$

and

$$I = \begin{Bmatrix} I_1^r \\ I_2^r \\ \vdots \\ I_n^r \end{Bmatrix} = \{I_1^c\ I_2^c\ \ldots\ I_n^c\},\quad I_i^r = (I_i^c)^t\ . \tag{2.41}$$

Using equations (2.39), (2.40) and (2.41), the constraint equation (2.37) can be written as n independent constraints as

$$A^t\ M_i^t + N_i^t = I_i^c\ ;\quad i = 1, 2, \ldots, n, \tag{2.42a}$$

or

$$M_i\ A + N_i = I_i^r\ ;\quad i = 1, 2, \ldots, n, \tag{2.42b}$$

The ith element of $\hat{X}$, from (2.38), is

$$\hat{x}_i = M_i \tilde{Y} + N_i \hat{X}_a \; ; \quad i = 1, 2, \ldots, n. \qquad (2.43)$$

Note that both eqns. (2.42) and (2.43) depend *only* upon the elements of the ith row M_i of M and the ith row N_i of N. Thus the ith variance (2.32) to be minimized is a function of the same n+m unknowns (the elements of M_i and N_i) as is the ith constraint, eqn. (2.42a) or (2.42b).

Substitution of (2.43) into (2.32) yields

$$\phi_i = E\,\{(M_i\tilde{Y} + N_i\hat{X}_a - x_i)^2\};$$
$$i = 1, 2, \ldots, n. \qquad (2.44)$$

Substitution of $\tilde{Y} = AX + V_y$ and $\hat{X}_a = X + V_x$ {from (2.29) and (2.30)} into (2.44) yields

$$\phi_i = E\,\{(M_iA + N_i)X + M_iV_y + N_iV_x - x_i\}^2$$
$$i = 1, 2, \ldots, n. \qquad (2.45)$$

Making use of (2.42b), (2.45) becomes

$$\phi_i = E\{(I_i^rX + M_iV_y + N_iV_x - x_i)^2\};$$
$$i = 1, 2, \ldots, n. \qquad (2.46)$$

Since $I_i^rX = x_i$, eqn. (2.46) reduces to

$$\phi_i = E\{(M_iV_y + N_iV_x)^2\}$$
$$i = 1, 2, \ldots, n. \qquad (2.47)$$

or

$$\phi_i = E\{(M_iV_y)^2 + 2(M_iV_y)\,(N_iV_x) + (N_iV_x)^2\}$$
$$i = 1, 2, \ldots, n \qquad (2.48)$$

which can be written as

$$\phi_i = E\{M_i(V_yV_y^t)M_i^t + 2M_i(V_yV_x^t)N_i^t + N_i(V_xV_x^t)N_i^t\};$$
$$i = 1, 2, \ldots, n. \qquad (2.49)$$

Since V_x and V_y are the only random variables in (2.49) and since

$$\Lambda_{yy} \equiv \Lambda_{vv} \equiv E(V_y V_y^t) \tag{2.50}$$

$$\Lambda_{xx_a} \equiv E(V_x V_x^t) \tag{2.51}$$

and by initial assumption

$$\Lambda_{xy} \equiv \Lambda_{vx} \equiv E(V_y \; V_x^t) = 0, \tag{2.52}$$

eqn. (2.49) becomes

$$\phi_i = M_i \Lambda_{yy} M_i^t + N_i \Lambda_{xx_a} N_i^t , \qquad i = 1, 2, \ldots, n. \tag{2.53}$$

The ith minimization problem can then be restated as: Determine M_i and N_i to minimize the ith eqn. of (2.53) subject to the ith constraint eqn. of (2.42).

Using the Method of Lagrange Multipliers (Appendix A), the augmented functions are defined as

$$\Phi_i = M_i \Lambda_{yy} M_i^t + N_i \Lambda_{xx_a} N_i^t + \lambda_i^t (I_i^c - A^t M_i^t - N_i^t) \; ; \qquad i = 1, 2, \ldots, n, \tag{2.54}$$

where

$$\lambda_i^t = (\lambda_{1_i} \lambda_{2_i} \ldots \lambda_{n_i}) \tag{2.55}$$

is the ith matrix of n Lagrange multipliers.

The necessary conditions for a minimum of (2.54) are

$$\nabla_{M_i^t} \Phi = 2\Lambda_{yy} M_i^t - A\lambda_i = 0 \; ; \qquad i = 1, 2, \ldots, n \tag{2.56}$$

$$\nabla_{N_i^t} \Phi = 2\Lambda_{xx_a} N_i^t - \lambda_i = 0 \qquad i = 1, 2, \ldots, n \tag{2.57}$$

and

$$I_i^c - A^t M_i^t - N_i^t = 0 \; ; \qquad i = 1, 2, \ldots, n \tag{2.58}$$

From eqns. (2.56) and (2.57), we obtain

$$M_i = \tfrac{1}{2}\lambda_i^t A^t \Lambda_{yy}^{-1} \; ; \; M_i^t = \tfrac{1}{2}\Lambda_{yy}^{-1} A \lambda_i \; ; \qquad i = 1, 2, \ldots n \quad (2.59)$$

and

$$N_i = \tfrac{1}{2}\lambda_i^t \Lambda_{xx_a}^{-1} \; ; \; N_i^t = \tfrac{1}{2}\Lambda_{xx_a}^{-1} \lambda_i \; ; \qquad i = 1, 2, \ldots n. \quad (2.60)$$

Substitution of (2.59) and (2.60) into (2.58) allows immediate solution for λ_i^t as

$$\lambda_i^t = 2I_i^r (A^t \Lambda_{yy}^{-1} A + \Lambda_{xx_a}^{-1})^{-1} \; ; \qquad i = 1, 2, \ldots n. \quad (2.61)$$

Then substituting (2.61) into (2.59) and (2.60), the rows of M and N are

$$M_i = I_i^r (A^t \Lambda_{yy}^{-1} A + \Lambda_{xx_a}^{-1})^{-1} A^t \Lambda_{yy}^{-1} \; ; \qquad i = 1, 2, \ldots n \quad (2.62)$$

and

$$N_i = I_i^r (A^t \Lambda_{yy}^{-1} A + \Lambda_{xx_a}^{-1})^{-1} \Lambda_{xx_a}^{-1} \; ; \qquad i = 1, 2, \ldots, n. \quad (2.63)$$

Therefore the M and N matrices are

$$M = (A^t \Lambda_{yy}^{-1} A + \Lambda_{xx_a}^{-1})^{-1} A^t \Lambda_{yy}^{-1} \quad (2.64)$$

and

$$N = (A^t \Lambda_{yy}^{-1} A + \Lambda_{xx_a}^{-1})^{-1} \Lambda_{xx_a}^{-1} \; . \quad (2.65)$$

Finally, substitution of (2.64) and (2.65) into (2.38) yields the Minimal Variance Estimator

$$\hat{X} = (A^t \Lambda_{yy}^{-1} A + \Lambda_{xx_a}^{-1})^{-1} (A^t \Lambda_{yy}^{-1} \tilde{Y} + \Lambda_{xx_a}^{-1} \hat{X}_a) \quad (2.66)$$

which allows rigorous processing of apriori state estimates $\hat{X}_a$ and associated covariance matrices Λ_{xx_a}.

Notice the following limiting cases:

(1) Apriori knowledge very poor

$(\Lambda_{xx_a} \to \infty,\ \Lambda_{xx_a}^{-1} \to 0)$,

then (2.66) reduces immediately to the standard minimal variance normal equations (2.28).

(2) Measurements very poor

$(\Lambda_{yy}^{-1} \to 0)$

then (2.66) yields $\hat{X} = \hat{X}_a$, an intuitively pleasing result!

Notice also that (2.66) can be obtained from the sequential least squares formulation of §1.4 by processing the apriori state information as a subset of "observations" as follows:

In eqns. (1.75) and (1.76) of the sequential estimation developments:

(1) Set k = 1 and ignore "1" subscript

(2) Set $A_2 = I$
$W_2 = \Lambda_{xx}^{-1}$
$Y_2 = \hat{X}_a$

Then one immediately obtains (2.66).

We thus conclude that the minimal variance estimate (2.66) is in all respects consistent with the sequential estimation results of §1.4; to start the sequential process, one would probably employ the apriori estimates as

$$X_1 = \hat{X}_a$$

$$P_1 = \Lambda_{xx_a},$$

and process subsequent measurement subsets $\{Y_k, A_k, W_k\}$ with $W_k = \Lambda_{yy}^{-1}$ for minimum variance estimates of X.

2.4 *Covariance Propagation in Linear Estimation Algorithms*

An important question which arises in most applications is the manner in which measurement uncertainty is propagated through the estimation algorithm into the estimated parameters. In Appendix B, we establish the following result: Given a linear matrix equation of the form

$$Z = M\, Y \tag{2.67}$$

where it is known that

$$E(Y) = \bar{Y} = \text{mean of } Y, \tag{2.68}$$

$$E((Y-\bar{Y})(Y-\bar{Y})^t) = \Lambda_{yy} = Y \text{ covariance matrix}, \tag{2.69}$$

and M is a known matrix, then the Z-covariance matrix is given by the transformation

$$\Lambda_{zz} \equiv E((Z-\bar{Z})\ (Z-\bar{Z})^t) = M\Lambda_{yy}M^t \tag{2.70}$$

Now, the weighted least square estimate is given explicitely by the normal equations

$$\hat{X} = (A^tWA)^{-1}\ A^tW\tilde{Y} \tag{1.30}$$

Comparing (2.67) and (1.30), we conclude by analogy with (2.70) that the $\hat{X}$ covariance matrix is given by

$$P = (A^tWA)^{-1}\ A^tW\Lambda_{yy}\ WA(A^tWA)^{-1}\ . \tag{2.71}$$

For the usual circumstance that the weight matrix is taken as the inverse of the measurement error covariance matrix (so that the least squares estimate is also a minimal variance estimate), the $\hat{X}$ covariance matrix (2.71) reduces as follows

$$P = (A^t\Lambda_{yy}^{-1}A)^{-1}\ \underbrace{A^t\Lambda_{yy}^{-1}\ \underbrace{\Lambda_{yy}\ \Lambda_{yy}^{-1}}_{I}\ A(A^t\Lambda_{yy}^{-1}A)^{-1}}_{I} \tag{2.72}$$

or

$$P = (A^t\Lambda_{yy}^{-1}A)^{-1} \tag{2.73}$$

For the case that an apriori estimate $\hat{X}_a$ and associated covariance P_a are available, it follows from eqn. (2.66) and (2.70), after some algebra, that given new data $\tilde{Y}$ and Λ_{yy}, the covariance matrix is given by

$$P = (P_a^{-1} + A^t\Lambda_{yy}A)^{-1} \tag{2.74}$$

Eqn. (2.74) can be alternatively written (see eqns. (1.78) and (1.93) as

$$P = P_a - P_aA^t(\Lambda_{yy} + AP_aA^t)^{-1}AP_a. \tag{2.75}$$

2.5 *Nonuniqueness of the Weight Matrix*

Here we study the truth that more than one weight matrix in the normal equations can yield identical X estimates. Actually two classes of weight matrices (which preserve $\hat{X}$) exist, the first is rather well known, the second is less known and its implications are more subtle.

We first consider the class of weight matrices which is formed by multiplying all elements of W by the same scalar α as

$$W' = \alpha W \tag{2.76}$$

The X estimate corresponding to W' follows from (1.30) as

$$\hat{X}' = (A^tW'A)^{-1}\ A^tW'\tilde{Y} \tag{2.77}$$

which, upon substituting (2.76) reduces as follows

$$\hat{X}' = \frac{1}{\alpha}(A^tWA)^{-1}A^t(\alpha W)\tilde{Y} = (A^tWA)^{-1}A^tWY$$

to

$$\hat{X}' \equiv \hat{X}.$$

Therefore, scaling all elements of W does not (formally) affect the estimate solution $\hat{X}$. Numerically, possibly significant errors may result if extremely small or extremely large values of α are used, due to computer truncation errors.

We now consider a second class of weight matrices obtained by adding a non-zero m x m matrix ΔW to W as

$$W'' = W + \Delta W \tag{2.78}$$

Then the estimate solution $\hat{X}''$ corresponding to W'' is obtained from eqn. (1.30) as

$$\hat{X}'' = (A^t W'' A)^{-1} A^t W'' \tilde{Y} . \tag{2.79}$$

Substitution of (2.78) into (2.79) yields

$$\hat{X}'' = \{A^t WA + (A^t \Delta W)A\}^{-1} \{A^t W\tilde{Y} + (A^t \Delta W)\tilde{Y}\} . \tag{2.80}$$

If $\Delta W \neq 0$ exists such that

$$A^t \Delta W = 0 , \tag{2.81}$$

Then equation (2.80) clearly reduces to

$$\hat{X}'' = (A^t WA)^{-1} A^t W\tilde{Y} \equiv \hat{X} .$$

There are, in fact, an infinity of matrices ΔW satisfying the *orthogonality constraint* (2.81). To see this, assume that all elements of ΔW except those in the first column are zero, then (2.81) becomes

$$A^t \Delta W = \begin{pmatrix} a_{11} & \cdots & a_{m1} \\ \vdots & \ddots & \vdots \\ a_{1n} & \cdots & a_{mn} \end{pmatrix} \begin{pmatrix} \Delta w_{11} & & \text{sym} \\ \Delta w_{12} & 0 & \\ \vdots & \vdots & \ddots \\ \Delta w_{1m} & 0 \cdots & 0 \end{pmatrix} = \begin{pmatrix} 0 & \cdots & 0 \\ \vdots & \ddots & \vdots \\ 0 & \cdots & 0 \end{pmatrix} \tag{2.82}$$

which yields the scalar equations

$$a_{1i}\, \Delta w_{11} + a_{2i}\, \Delta w_{12} + \ldots + a_{mi}\, \Delta w_{1m} = 0 \; ; \quad i = 1, 2, \ldots, n \tag{2.83a}$$

and

$$a_{1i} \Delta w_{1j} = 0 \; ; \quad i = 2, 3, \ldots n \qquad j = 2, 3, \ldots n \tag{2.83b}$$

Ignoring the case that the a_{1i}'s are zero and the trivial case that the Δw_{1j}'s are zero, then equations (2.83a) provide n equations.

to be satisfied by the remaining m unspecified Δw_{1j}'s. Since any n of the Δw_{1j}'s can be determined to satisfy (2.83), while the remaining m − n Δw_{1j}'s can be given arbitrary values, it follows that an infinity of ΔW matrices satisfy (2.82) and therefore (2.81).

The fact that more than one weight matrix yields the same X-estimates is no cause for alarm. Interpreting the covariance matrix as the inverse of the measurement error covariance matrix associated with a specific $\tilde{Y}$ of measurements, the above results imply that one can obtain the same X-estimate from given measured Y-values, for a variety of measurement weights, according to (2.76) or (2.78) and (2.81). A most interesting question can be asked regarding the covariance matrix of the estimated parameters. From equation (2.73), we established that the estimated parameter covariance matrix is

$$\Lambda_{xx} = (A^t W A)^{-1} \ , \ W = \Lambda_{yy}^{-1} \ . \tag{2.84}$$

For the first class of weight matrices $W' = \alpha W$ note that

$$\Lambda'_{xx} = \frac{1}{\alpha} (A^t W A)^{-1} = \frac{1}{\alpha} (A^t \Lambda_{yy}^{-1} A)^{-1}, \tag{2.85}$$

or

$$\Lambda'_{xx} = \frac{1}{\alpha} \Lambda_{xx} \ . \tag{2.86}$$

Thus linear scaling of the observation weight matrix results in reciprocal linear scaling of the estimated parameter covariance matrix, an intuitively reasonable result.

Considering now the second class of error covariance matrices $W'' = W + \Delta W$, with $A^t \ \Delta W = 0$, it follows from (2.84) that

$$\Lambda''_{xx} = (A^t \ WA + A^t \ \Delta WA)^{-1} = (A^t WA)^{-1} \tag{2.87}$$

or

$$\Lambda''_{xx} = \Lambda_{xx}. \tag{2.88}$$

Thus the additive class of observation weight matrices preserves not only the X-estimates, but also the associated covariance matrix. It may prove possible, in some applications, to exploit this truth since a family of measurement variances can result in the same estimates and associated uncertainties.

The above properties are illustrated in the following numerical example.

Example 2.1

Given the linear system

$$Y = AX$$

where

$$\tilde{Y} = \begin{Bmatrix} 2 \\ 1 \\ 3 \end{Bmatrix}$$

$$A = \begin{bmatrix} 1 & 3 \\ 2 & 2 \\ 3 & 4 \end{bmatrix}$$

For each of the three weight matrices

$$W = \begin{bmatrix} 1 & 0 & 0 \\ 0 & 1 & 0 \\ 0 & 0 & 1 \end{bmatrix} ,$$

$$W' = 3W ,$$

and

$$W'' = W + \begin{bmatrix} 1/4 & 5/8 & -1/2 \\ 5/8 & 25/16 & -5/4 \\ -1/2 & -5/4 & 1 \end{bmatrix} ,$$

determine the least square estimates

$$\hat{X} = (A^tWA)^{-1}A^tW\tilde{Y}$$

$$\hat{X}' = (A^tW'A)^{-1}\ A^tW'\tilde{Y}$$

$$\hat{X}'' = (A^tW''A)^{-1}\ A^tW''\tilde{Y}$$

and the corresponding covariance estimates

$$\Lambda_{xx} = (A^tWA)^{-1},\ \Lambda'_{xx} = (A^tW'A)^{-1},\ \text{and}\ \Lambda''_{xx} = (A^tW''A)^{-1}$$

the reader can verify the numerical results

$$\hat{X} = \hat{X}' = \hat{X}'' = \begin{Bmatrix} -1/15 \\ 11/15 \end{Bmatrix}$$

and

$$\Lambda_{xx} \equiv \Lambda''_{xx} = \begin{bmatrix} 29/45 & -19/45 \\ -19/45 & 14/45 \end{bmatrix}$$

$$\Lambda'_{xx} \equiv \frac{1}{3}\,\Lambda_{xx} = \begin{bmatrix} 29/135 & -19/135 \\ -19/135 & 14/135 \end{bmatrix} \quad .$$

2.6 Remarks

In this chapter, we have presented a minimal variance approach to establish a class of linear estimation algorithms, and developed certain important properties of the weight-matrix. The end products of the minimal variance developments of §2.2 and 2.3 are seen to be equivalent to the linear least squares results of §1.3 - 1.4, with the interpretation of the weight matrix as the measurement error covariance matrix, and interpretation of apriori parameter estimates as a measurement subset (in the sequential least squares developments of §1.4). The minimal variance discussion of §2.2 and 2.3 were motivated by Gura's 1967) excellent discussion. It should be noted that several alternative probabilistic paths (e.g., "Baye's criterion" (ref. 2.2, 2.3, 2.4), "maximum liklihood criterion" (ref. 2.5, 2.6)) lead, for most applications, to essentially the same algorithms as the least squares and minimal variance criteria. Each of these paths provides certain illuminations and useful insights, however; the constraints on the level and length of the present monograph preclude their inclusion here.

The discussion in §2.5 on nonuniqueness of the weight matrix was motivated by Cohen's (ref. 2.7) lecture notes on the subject.

It should be noted that specification and calculations involving the weight matrices are the source of most practical difficulties encountered in applications.

2.7 Exercises for Chapter 2

2.1 Write a simple computer program to simulate measurements of some discretely measured process

$$\tilde{y}_j = x_1 + x_2 \sin 10t_j + x_3\, e^{2t_j^2} + v_j, \quad j = 1,2,\cdots,11 \tag{2.89}$$

$$\{t_j\} = \{0,\ 0.1,\ 0.2,\cdots,\ 1.0\}$$

where the true $(x_1,\ x_2,\ x_3)$ are $(1,\ 1,\ 1)$ and the measurement errors are synthetic Gaussian (normal) random numbers with zero mean. If your computer does not have a Gaussian random number generator, use the simple generator*

```
Z = 0
FOR I = 1 to 12
Z = Z + RND(I)                                  (2.90)
NEXT I
V = ZSIGMA * (Z - 6) + ZBAR
```

The above algorithm requires that your computer have a library subroutine RND which will yield uniformly distributed random numbers from 0 to 1; the output (V) is a pseudo-Gaussian number with mean = ZBAR and standard deviation = $(\text{variance})^{\frac{1}{2}}$ = ZSIGMA. Clearly the algorithm (2.90) must be executed once to determine v_j for each of the 11 measurements in (2.89).

You are given the measurement error covariance matrix

$$\Lambda_{yy} = E(VV^t) = \begin{pmatrix} (\text{SIGMA}_1)^2 & & & \\ & (\text{SIGMA}_2)^2 & & \\ & & \ddots & \\ & & & (\text{SIGMA}_{11})^2 \end{pmatrix} \tag{2.91}$$

---

*This algorithm is based upon the truth (central limit theorem) that the sum of random variables belonging to arbitrary symmetric distributions approach a Gaussian distribution.

where,

$$\begin{array}{llll} \text{SIGMA}_1 = 0.001 & \text{SIGMA}_7 = 0.010 \\ \text{SIGMA}_2 = 0.002 & \text{SIGMA}_8 = 0.007 \\ \text{SIGMA}_3 = 0.005 & \text{SIGMA}_9 = 0.020 \\ \text{SIGMA}_4 = 0.010 & \text{SIGMA}_{10} = 0.006 \\ \text{SIGMA}_5 = 0.008 & \text{SIGMA}_{11} = 0.001 \\ \text{SIGMA}_6 = 0.002 & \end{array} \tag{2.92}$$

and required to use (2.90) and (2.89) to simulate measurements $\{\tilde{y}_1, \tilde{y}_2, \ldots, \tilde{y}_{11}\}$. You are also given the apriori X-estimates

$$\hat{X}_a^t = \{\hat{X}_1 \hat{X}_2 \hat{X}_3\} = \{1.01 \quad 0.98 \quad 0.99\} \tag{2.93}$$

and associated apriori covariance matrix

$$\Lambda_{xx_a} = \begin{pmatrix} 0.001 & 0 & 0 \\ 0 & 0.001 & 0 \\ 0 & 0 & 0.001 \end{pmatrix} . \tag{2.94}$$

Your tasks are the following:

(A) Use the minimal variance estimation version of the normal equations

$$\hat{X} = P(A^t \Lambda_{yy}^{-1} \tilde{Y} + \Lambda_{xx_a}^{-1} \hat{X}_a) \tag{2.66}$$

to compute the parameter estimates and estimate covariance matrix

$$P = (A^t \Lambda_{yy}^{-1} A + \Lambda_{xx}^{-1})^{-1}. \tag{2.74}$$

(observe that the jth row of A is $(1 \ \sin 10t_j \ e^{2t^2_j})$).

Calculate the mean and standard deviation of the residual

$$r_j = \tilde{y}_j - (\hat{x}_1 + \hat{x}_2 \sin 10t_j + \hat{x}_3 e^{2t_j^2})$$

as

$$r = \frac{1}{11} \sum_{j=1}^{11} r_j$$

$$\sigma_r = \{\frac{1}{10} \sum_{j=1}^{11} r_j^2\}^{\frac{1}{2}} \tag{2.95}$$

(B) Do a parametric study in which you hold the apriori estimate covariance Λ_{xx_a} fixed, but vary the measurement error covariance according to

$$\Lambda'_{yy} = \alpha \Lambda_{yy} \tag{2.78}$$

with

$$\alpha = 10^{-3}, 10^{-2}, 10^{-1}, 10, 10^2, 10^3 \quad .$$

and study the behavior of the calculated results from eqns. (2.66), (2.74), (2.93), (2.94) and (2.95).

(C) Do a parametric study in which Λ_{yy} is held fixed, but Λ_{xx_a} is varied according to

$$\Lambda''_{xx_a} = \alpha \Lambda_{xx_a} \tag{2.79}$$

with α taking on the same values as in (B). Compare the results from eqns. (2.66), (2.74), (2.93), (2.94) and (2.95) with those of part (B).

2.2 A "Monte Carlo approach" to calculating covariance matrices is often necessary for nonlinear problems. The algorithm has the following structure: Given a functional dependence of two sets of random variables in the form

$$z_i = F_i(y_1, y_2, \ldots, y_m) \qquad i = 1, 2, \ldots, n \tag{2.96}$$

where the y_j are random variables whose joint probability density function is known; the F_i in (2.96) are generally nonlinear functions. The Monte Carlo approach requires that the probability density function of y_j be sampled many times and equations (2.96) used to calculate corresponding samples of the z_i joint distribution. Thus if the kth particular sample ("simulated measurement") of the y_j values is denoted as

$$\{\tilde{y}_{1k}, \tilde{y}_{2k}, \cdots, \tilde{y}_{mk}\} \; ; \quad k = 1,2,\cdots,M \tag{2.97}$$

Then the corresponding z_i sample is calculated from (2.96) as

$$z_{ik} = F_i(y_{1k}, y_{2k}, \cdots, y_{mk}) \; ; \quad k = 1,2,\cdots M. \tag{2.98}$$

The first two moments of z_i's joint density function are then approximated as

$$\bar{z}_i = E(z_i) \simeq \frac{1}{M} \sum_{k=1}^{M} z_{ik} \tag{2.99}$$

and

$$\Lambda_{zz} = E\{(Z-\bar{Z})\;(Z-\bar{Z})^t\}$$

$$\simeq \frac{1}{M-1} \sum_{k=1}^{M} \begin{pmatrix} (z_{1k}-\bar{z}_1)^2 & (z_{1k}-\bar{z}_1)(z_{2k}-\bar{z}_2) \cdots & (z_{1k}-\bar{z}_1)(z_{nk}-\bar{z}_n) \\ & (z_{2k}-\bar{z}_2)^2 & \cdots (z_{2k}-\bar{z}_2)(z_{nk}-\bar{z}_n) \\ & & \ddots \quad \vdots \\ \text{Symmetric} & & (z_{nk}-\bar{z}_n)^2 \end{pmatrix} \tag{2.100}$$

with

$$\bar{Z}^t \equiv \{\bar{z}_1 \cdots \bar{z}_n\}.$$

The above Monte Carlo approach can be used to experimentally verify the interpretation of $P = (A^t \Lambda_{yy}^{-1} A)^{-1}$ as the $\hat{X}$ covariance matrix in the minimal variance estimate

$$\hat{X} = P\, A^t \Lambda_{yy}^{-1} \tilde{Y} \; . \tag{2.28}$$

To carry out this experiment, use the model (2.89) and the random number generator (2.90) to simulate M = 100 sets of y-measurements. For each set (e.g., the kth) of measurements, the corresponding $\hat{X}_k$ follows from (2.28) as

$$\hat{X}_k = P\, A^t\, \Lambda_{yy}^{-1}\, \tilde{Y}_k \tag{2.101}$$

Then the $\hat{X}$ mean and covariance matrices can be approximated according to (2.99) and (2.100) as

$$E(\hat{X}) \simeq \frac{1}{M} \sum_{k=1}^{M} \hat{X}_k \tag{2.102}$$

$$\begin{aligned} \Lambda_{xx} &\equiv E\{[\hat{X} - E(\hat{X})]\,[\hat{X} - E(\hat{X})]^t\} \\ &\simeq \frac{1}{M-1} \sum_{k=1}^{M} \{[\hat{X}_k - E(\hat{X})]\,[\hat{X}_k - E(\hat{X})]^t\}, \end{aligned} \tag{2.103}$$

for M = 100.

Equation (2.103) should be compared element-by-element with $P = (A^t\, \Lambda_{yy}^{-1}\, A)^{-1}$, whereas eqn. (2.102) should compare favorably with the true values $X^t = (1 \;\; 1 \;\; 1)$.

2.8 *References for Chapter 2*

2.1 Gura, I.A., "An Algebraic Approach to Optimal State Estimation", Hughes Aircraft Missile and Space Systems Division Report No. SSD70072R, (1967), El Segundo, Ca.

2.2 Ho, Y.C., and R.C.K. Lee, "A Bayesian Approach to Problems in Stochastic Estimation and Control", IEEE Trans., PGAC, AC-9 (Oct. 1964), pp 333-339.

2.3 Pugachev, V.S., "The Method of Determining Optimum Systems Using Bayes Criteria", IRE Trans. PGCT, CT-7 (Dec. 1960) pp 491-505.

2.4 Sage, A.P. and J.L. Melsa, Estimation Theory, McGraw Hill (1971), pp 175-237.

2.5 Solloway, C.B., "Elements of Orbit Determination", Jet Propulsion Laboratory, Engineering Planning Document Report # 255, Pasadena, Ca. (1964).

2.6 Richards, F.S.G., "A Method of Maximum-Liklihood Estimation", J. Royal Statistical Soc., Ser. B., 23 (1961), pp 469-475.

2.7 Cohen, C., "Lectures on Least Squares", edited by T.M. Alexander, U.S. Naval Weapons Laboratory Report KGR-M1, Dahlgren, Va., May 1972.

3

Parameter Estimation: Applications

3.1 Preliminary Comments

In this chapter, seven example applications are presented in which the methods of the first two chapters can be used to advantage. Exclusive of the elementary examples of §3.2, the problems and solutions are idealizations of "real world" applications which are documented in the literature cited. In several instances, the system models used in the analysis are stated but not formulated from first principles, due to the space limitations here. However, care has been exercised to employ examples whose mathematical models are either self-evident, or are well-documented in the cited literature.

Applications of estimation algorithms to these relatively transparent mathematical models should prove most useful in developing the reader's appreciation for the role of the foregoing developments in the solution of practical problems.*

*Unless otherwise noted, all calculations discussed in this chapter were performed on the University of Virginia'a Control Data Corporation CYBER 172 computer system, using single precision (60 bit word length). Fortran was used throughout; the programs for the more elaborate applications are documented in the reports cited.

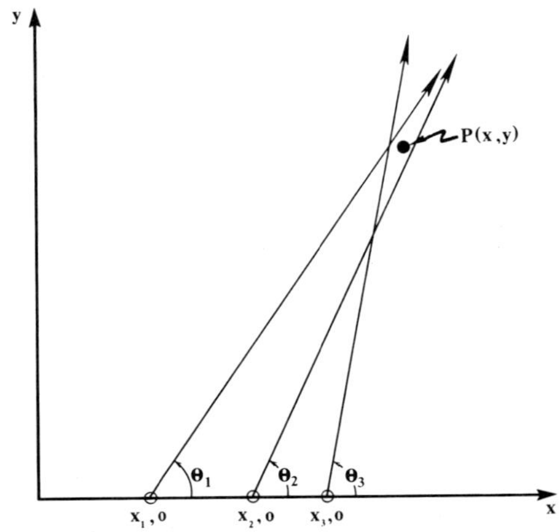

Figure 3.1 Planar Triangulation from Known Base Points

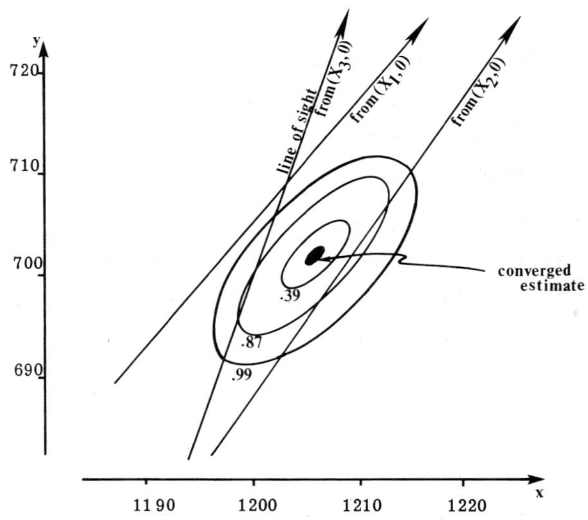

Figure 3.2 Confidence Ellipses for the Gaussian Least Square Differential Correction Solution of Problem 3.2.1

3.2 Planar Triangulation

*Example Problem 3.2.1 Planar Triangulation From Known Base Points**

With reference to Figure 3.1, suppose that a surveyor has collected data to estimate the location (x,y) of a point P. Suppose that the measurements consist of the azimuth θ of P from several perfectly known points along a baseline (the x-axis). P is assumed, for simplicity, to lie in the x-y plane and the measurement errors are assumed to be uncorrelated.

Given the measurements

j	x_j(ft)	$\tilde{\theta}_j$(deg)	$\sigma^2_{\theta j}$(deg^2)
1	0	30.1	0.01
2	500	45.0	0.01
3	1000	73.6	0.04

The objective is to determine the optimal least square (also minimal variance) estimates of the coordinates (x,y) of P and the associated covariance matrix.

Solution 3.2.1A Gaussian Least Square Differential Correction

From the geometry of Figure 3.1, it follows that perfect measurements θ_j from base points $(x_j,0)$ satisfy the equations

$$\theta_j = F_j(x,y) \equiv \tan^{-1}\left(\frac{y}{x-x_j}\right), \quad j = 1,2,3. \tag{3.1}$$

The objective is to determine estimates (x,y) which minimize the residuals' weighted sum of squares

$$\phi = \sum_{j=1}^{3} \{\tilde{\theta}_j - F_j(\hat{x},\hat{y})\}^2 \frac{1}{\sigma^2_{\theta_j}} . \tag{3.2}$$

The Gaussian least square differential correction algorithm (§1.5) for successively improving given initial approximations (x_c,y_c) of $(\hat{x},\hat{y})$ is

---

*This problem was motivated by an example in Bryson and Ho, *Applied Optimal Control*, pp 352-355,

$$\begin{Bmatrix} x_c \\ y_c \end{Bmatrix}^{(k+1)} = \begin{Bmatrix} x_c \\ y_c \end{Bmatrix}^{(k)} + \{(A^tWA)^{-1}\ A^tW\Delta Y\}^{(k)} \tag{3.3}$$

where, in this case, the matrices are defined as

$$\Delta Y = \begin{Bmatrix} \tilde{\theta}_1 - F_1(x_c, y_c) \\ \tilde{\theta}_2 - F_2(x_c, y_c) \\ \tilde{\theta}_3 - F_3(x_c, y_c) \end{Bmatrix} \tag{3.4}$$

$$A = \begin{bmatrix} \frac{\partial F_1}{\partial x_c} & \frac{\partial F_1}{\partial y_c} \\ \frac{\partial F_2}{\partial x_c} & \frac{\partial F_2}{\partial y_c} \\ \frac{\partial F_3}{\partial x_c} & \frac{\partial F_3}{\partial y_c} \end{bmatrix} \tag{3.5}$$

and

$$W = \begin{bmatrix} \frac{1}{\sigma^2_{\theta_1}} & & 0 \\ & \frac{1}{\sigma^2_{\theta_2}} & \\ 0 & & \frac{1}{\sigma^2_{\theta_3}} \end{bmatrix}. \tag{3.6}$$

Given below in Table 3.1 are the results from the first three iterations of the above algorithm, given the initial starting estimates ($x_c \equiv 1210$ ft and $y_c = 700$ ft) of the point P's location.

The covariance matrix associated with the converged estimates can be approximated using the results of §2.4 as

$$\Lambda \equiv E \begin{pmatrix} (x-\hat{x})^2 & (x-\hat{x})(y-\hat{y}) \\ (y-\hat{y})(x-\hat{x}) & (y-\hat{y})^2 \end{pmatrix} \cong (A^tWA)^{-1} = \begin{pmatrix} 10.093 & 10.114 \\ 10.114 & 12.595 \end{pmatrix} \tag{3.7}$$

TABLE 3.1

ITERATIVE SOLUTION OF PROBLEM 3.2.1 USING GAUSSIAN LEAST SQUARE DIFFERENTIAL CORRECTION

Correction Number	$x_c = x$	$y_c = y$	ϕ_c Sum Square of Residuals (Eqn. 3.2)
0	1210.000	700.000	19.0013
1	1204.800	701.804	3.3893
2	1204.783	701.774	3.3892
3	1204.783	701.775	3.3892

TABLE 3.2

GRADIENT SOLUTION OF PROBLEM 3.2.1

Iteration Number			
0	1210.00	700.00	19.00
1	1202.46	706.57	18.48
2	1206.05	703.09	3.54
⋮			
9	1204.86	701.91	3.39
10	1204.90	701.84	3.39
⋮			
18	1204.79	701.79	3.39

Equation (3.7) is valid to the approximation that a linearized prediction of θ's for dispersions about (x,y) is valid everywhere within (say) the 99% confidence ellipse.

Assuming that the observation errors are normally distributed and that the final linearization is valid, then the dispersions of x and y about their true values will also be normally distributed. For a joint normal density function, the surfaces of constant probability in 2-space are ellipses; these ellipses are a convenient device for visualizing uncertainty in the converged estimates. Preparing to make use of the developments in Appendix B, we first determine the eigenvalues and eigenvectors of the covariance matrix (3.7) to be

$$\sigma_1^2 = 21.94 \text{ ft}^2 , \quad e_1 = \begin{Bmatrix} 1.000 \\ 1.083 \end{Bmatrix}$$

$$\sigma_2^2 = 1.64 \text{ ft}^2 , \quad e_2 = \begin{Bmatrix} 1.000 \\ -0.923 \end{Bmatrix}$$

The eigenvectors define orientation of the principal axes of the ellipses of constant probability, and the square roots of the eigenvalues are the respective 1σ semi-axes.

The 1-, 2-, and 3-sigma likelihood ellipses are sketched in Figure 3.2. The probability that the true P(x,y) lies within these ellipses is .39, .87, and .99, respectively.

Solution 3.2.1 B Method of Gradients

As an alternate to the above Gaussian differential correction solution, the sum of squares of the residuals can be minimized by the method of gradients (or "method of steepest descent" see Appendix A). The essential difference between the two approaches is that the gradient corrections are based upon linearizing the prediction of the sum square of the residuals (3.2), whereas, the Gaussian differential correction algorithm is based upon linearizing the equations which model the measurements (3.1).

The method of gradients algorithm for the problem at hand is summarized as follows:

- Given estimates $X^{(0)} = \begin{Bmatrix} x_c \\ y_c \end{Bmatrix}^{(0)}$
- ✶ $k = k + 1$
- Compute $\phi^{(k)}$ (9.2) and $G^{(k)} = \begin{Bmatrix} \dfrac{\partial\phi}{\partial x_c} \\ \dfrac{\partial\phi}{\partial y_c} \end{Bmatrix}^{(k)}$ (3.9)

* Conduct one-variable search to find a scalar α such that

$$\phi^{(k)} = \phi(X^{(k)} - \alpha G^{(k)}) \tag{3.10}$$

is minimized.

* Apply corrections

$$X^{(k+1)} = \begin{Bmatrix} x_c \\ y_c \end{Bmatrix}^{(k+1)} = \begin{Bmatrix} x_c \\ y_c \end{Bmatrix}^{(k)} - \alpha G^{(k)} \tag{3.11}$$

* Return to step ✱.

The numerical results of 18 iterations of the above algorithm are given in Table 3.2. These iterations should be compared with the Gaussian differential correction results in Table 3.1. As might be anticipated, the Gaussian algorithm was more efficient here than the gradient algorithm in terms of the number of required iterations. Notice also that the Gaussian differential correction procedure provides uncertainty information (the estimate covariance matrix) as a by-product, whereas the gradient algorithm does not. Gradient algorithms typically exhibit inferior terminal convergence in comparison to Gaussian differential correction algorithms, unless A^tWA is poorly conditioned, in which case gradient-type algorithms provide attractive alternatives.

Example Problem 3.2.2 Planar Triangulation From Imperfectly Known Base Points

Problem §3.2.1 assumed optimistically that all aximuth measurements were made from perfectly known base points $(x_j, 0)$. This restriction is now removed and the Gaussian differential correction solution generalized to allow estimation of the measurement base point coordinates as well as the coordinates of P. The first measurement base point is adopted as the origin ($x_1 = y_1 = 0$) and the relative coordinates (x_2, y_2), (x_3, y_3) are admitted as four additional unknowns.

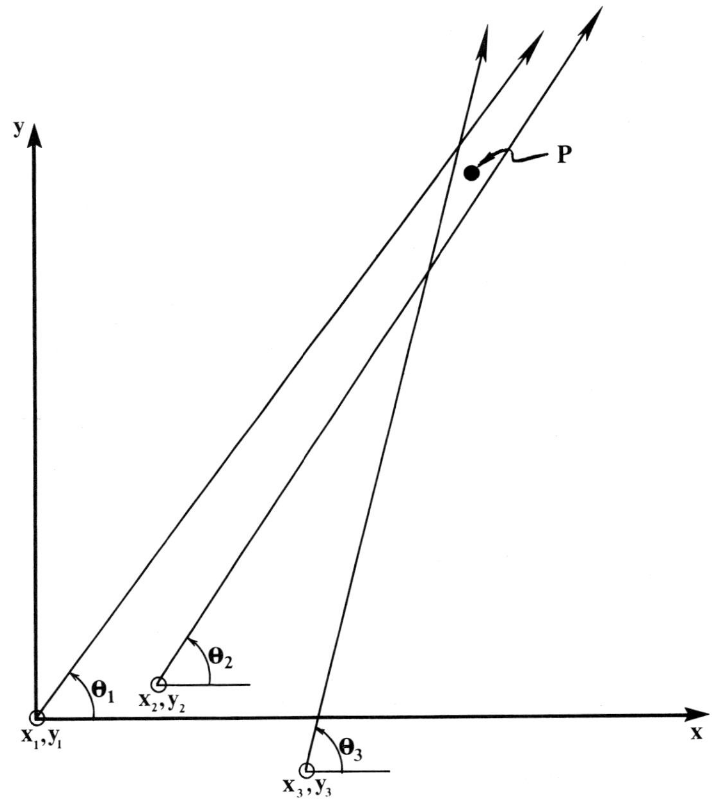

Figure 3.3 Planar Triangulation from Uncertain Base Points

Solution 3.2.1 Gaussian Least Square Differential Correction

The observations are modeled (refer to Figure 3.3) as

$$\tilde{\theta}_j = \tan^{-1}\left(\frac{y-y_j}{x-x_j}\right) + v_{\theta_j} \qquad j = 1,2,3 \tag{3.12}$$

$$\tilde{x}_j = x_j + v_{x_j} \qquad j = 2,3 \tag{3.13}$$

$$\tilde{y}_j = y_j + v_{y_j} \qquad j = 2,3 \tag{3.14}$$

Thus, there are seven "observable" parameters

$\{\theta_1,\theta_2,\theta_3,x_2,y_2,x_3,y_3\}$

and six unknown (to be estimated) "state" parameters

$\{x,y,x_2,y_2, x_3,y_3\}$.

The dual role of $\{x_2,y_2,x_3,y_3\}$ as "state" and "observable" parameters should present no particular conceptual difficulty if one recognizes that equations (3.13) and (3.14) are the simplest possible dependence of the observable parameters upon the unknown state variables.

Given the starting estimates

$$X_c^{(0)} \equiv \begin{bmatrix} x_c \\ y_c \\ x_{2c} \\ y_{2c} \\ x_{3c} \\ y_{4c} \end{bmatrix} = \begin{bmatrix} 1210 \\ 700 \\ 500 \\ 50 \\ 1000 \\ -100 \end{bmatrix} \tag{3.15}$$

and the observation in Table 3.3, the Gaussian differential correction algorithm for successive approximations is

$$X_c^{(k+1)} = X_c^{(k)} + \{(A^tWA)^{-1}A^tW\Delta Y\}^{(k)} \tag{3.16}$$

where

$$F_{jc} = \tan^{-1}\left(\frac{y_c - y_{jc}}{x_c - x_{jc}}\right) \tag{3.17}$$

$$Y = \begin{bmatrix} \tilde{\theta}_1 & - & F_{1c} \\ \tilde{\theta}_2 & - & F_{2c} \\ \tilde{\theta}_3 & - & F_{2c} \\ \tilde{x}_2 & - & x_{2c} \\ \tilde{y}_2 & - & y_{2c} \\ \tilde{x}_3 & - & x_{3c} \\ \tilde{y}_3 & - & y_{3c} \end{bmatrix} \tag{3.18}$$

TABLE 3.3

MEASURED DATA FOR PROBLEM 3.2.2

j	$\tilde{x}_j$	$\sigma^2_{x_j}$	$\tilde{y}_j$	$\sigma^2_{y_j}$	$\tilde{\theta}_j$	$\sigma^2_{\theta_j}$
1	0	0	0	0	30.1	0.01
2	500	100	50	144	45.0	0.01
3	1000	25	-100	100	73.6	0.04

TABLE 3.4

CONVERGENCE HISTORY FOR GAUSSIAN DIFFERENTIAL CORRECTION SOLUTION OF PROBLEM 3.2.2

Iteration Number	x	y	x_{2c}	y_{2c}	x_{3c}	y_{3c}
0	1200.00	700.00	500.00	50.00	1000.00	-100.00
1	1229.11	748.61	500.14	49.79	999.98	-99.98
2	1229.30	749.59	500.14	49.79	999.98	-99.98
3	1229.26	749.54	500.14	49.79	999.98	-99.98

$$A = \begin{bmatrix} \left.\frac{\partial F_1}{\partial x}\right|_c & \left.\frac{\partial F_1}{\partial y}\right|_c & 0 & 0 & 0 & 0 \\ \left.\frac{\partial F_2}{\partial x}\right|_c & \left.\frac{\partial F_2}{\partial y}\right|_c & \left.\frac{\partial F_2}{\partial x_2}\right|_c & \left.\frac{\partial F_2}{\partial y_2}\right|_c & 0 & 0 \\ \left.\frac{\partial F_3}{\partial x}\right|_c & \left.\frac{\partial F_3}{\partial y}\right|_c & 0 & 0 & \left.\frac{\partial F_3}{\partial x_3}\right|_c & \left.\frac{\partial F_3}{\partial y_3}\right|_c \\ 0 & 0 & -1 & 0 & 0 & 0 \\ 0 & 0 & 0 & -1 & 0 & 0 \\ 0 & 0 & 0 & 0 & -1 & 0 \\ 0 & 0 & 0 & 0 & 0 & -1 \end{bmatrix} \quad (3.19)$$

and

$$W^{-1} = \begin{bmatrix} \sigma^2_{\theta_1} & & & & & & 0 \\ & \sigma^2_{\theta_2} & & & & & \\ & & \sigma^2_{\theta_3} & & & & \\ & & & \sigma^2_{x_2} & & & \\ & & & & \sigma^2_{y_2} & & \\ & & & & & \sigma^2_{x_3} & \\ 0 & & & & & & \sigma^2_{y_3} \end{bmatrix} . \quad (3.20)$$

The convergence history resulting from successive approximations (3.16) is displayed in Table 3.4.

3.3 Stellar Resection Photogrammetry/Spacecraft Orientation Estimation

With reference to Figure 3.4, we consider the problem of determining the angular orientation of a space vehicle from photographs of the stars made from one or more spacecraft-fixed cameras. The brightest 150,000 stars' spherical coordinate angles (α, δ see figure 3.5) are available in a computer accesible catalog (ref. 3.2). Referring to figures 3.5 to 3.7, given the camera orientation angles (ϕ_1, ϕ_2, ϕ_3), it is established in ref. 3.3 that the photograph image plane coordinates of the jth star are determined by the stellar *colinearity equations*

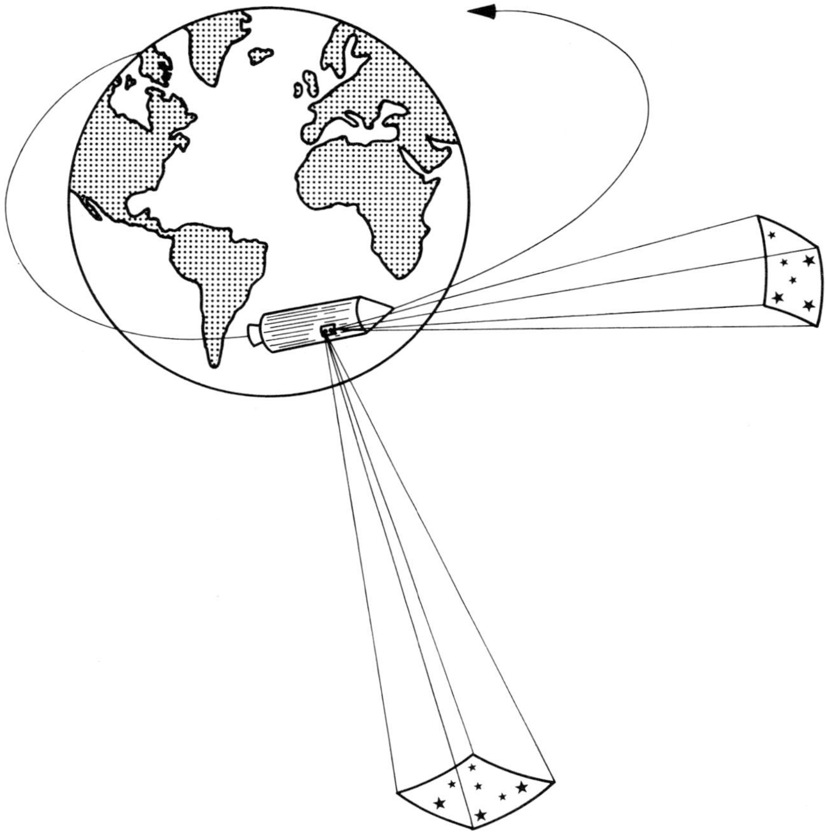

Figure 3. 4 Spacecraft Orientation Estimation from Star Photography[as an alternative to conventional photography, an electro-optical sensor(ref. 3. 3 and 3. 4) is under development which, in conjunction with an on-board computer, recognizes the star patterns and estimates spacecraft orientation in near-real-time]

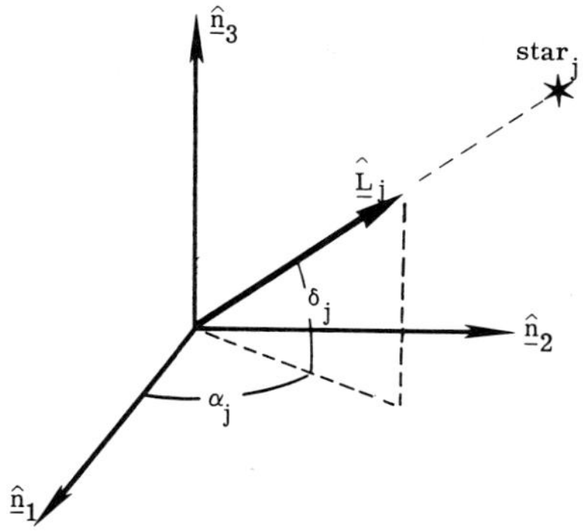

Figure 3.5 Spherical Coordinates α_j (right ascension) and δ_j (declination) Orienting the Line of Sight Vector $\hat{\underline{L}}_j$

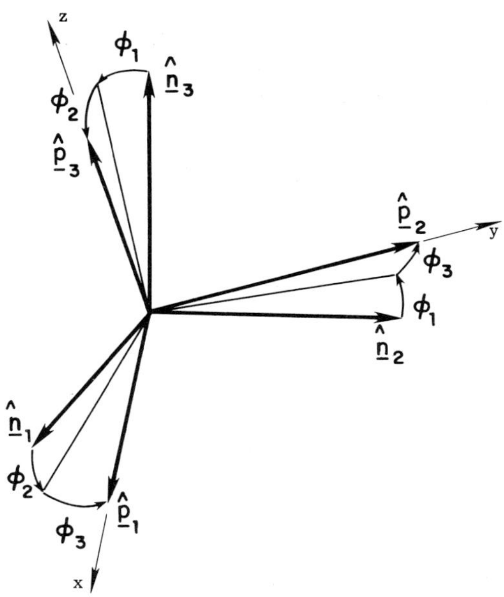

Figure 3.6 Orientation of the Photograph Coordinate System $\{\hat{\underline{p}}\}$ with respect to Inertial Space $\{\hat{\underline{n}}\}$

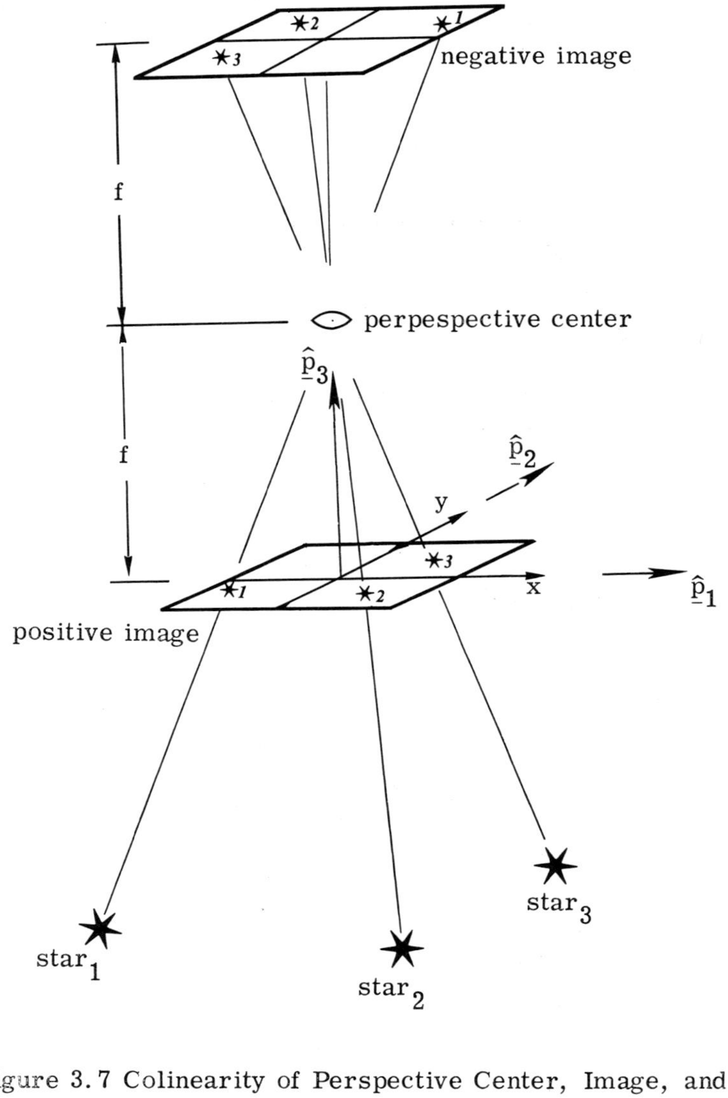

Figure 3.7 Colinearity of Perspective Center, Image, and Object

$$x_j = g_x(\alpha_j,\delta_j,\phi_1,\phi_2,\phi_3) = -f\left(\frac{c_{11}\,L_{xj} + c_{12}L_{yj} + c_{13}L_{zj}}{c_{31}\,L_{xj} + c_{32}L_{yj} + c_{33}L_{zj}}\right) \tag{3.21}$$

$$y_j = g_y(\alpha_j,\delta_j,\phi_1,\phi_2,\phi_3) = -f\left(\frac{c_{21}\,L_{xj} + c_{22}L_{yj} + c_{23}L_{zj}}{c_{31}\,L_{xj} + c_{32}L_{yj} + c_{33}L_{zj}}\right)$$

where the direction cosine matrix orienting the camera in inertial space is

$$\begin{pmatrix} c_{11} & c_{12} & c_{13} \\ c_{21} & c_{22} & c_{23} \\ c_{31} & c_{32} & c_{33} \end{pmatrix} = \begin{pmatrix} \cos\phi_3 & \sin\phi_3 & 0 \\ -\sin\phi_3 & \cos\phi_3 & 0 \\ 0 & 0 & 1 \end{pmatrix}\begin{pmatrix} \cos\phi_2 & 0 & -\sin\phi_2 \\ 0 & 1 & 0 \\ \sin\phi_2 & 0 & \cos\phi_2 \end{pmatrix}\begin{pmatrix} 1 & 0 & 0 \\ 0 & \cos\phi_1 & \sin\phi_1 \\ 0 & -\sin\phi_1 & \cos\phi_1 \end{pmatrix} \tag{3.22}$$

and the inertial components of the vector toward the jth star are

$$\begin{aligned} L_{xj} &= \cos\delta_j\ \cos\alpha_j \\ L_{yj} &= \cos\delta_j\ \sin\alpha_j \\ L_{zj} &= \sin\delta_j \end{aligned} \tag{3.23}$$

and the camera focal length f is known from apriori calibration.

Unfortunately, (ϕ_1,ϕ_2,ϕ_3) are usually poorly known, but if the measured stars can be identified* as specific cataloged stars, then predicted image locations can be calculated from the best apriori estimate of the camera orientation angles (ϕ_1,ϕ_2,ϕ_3), using eqns. (3.21). The predicted image coordinates will generally differ from their respective measured values; a nonlinear least square differential correction process can be structured to determine optimal estimates $(\hat{\phi}_1,\hat{\phi}_2,\hat{\phi}_3)$ which brings equations (3.21) into least square agreement with the measurements. To illustrate a successful application of this approach, consider the following problem.

---

*See ref. 3.4 for a pattern recognition technique which can be employed to automate the association of the measured images with the cataloged stars.

TABLE 3.5

SIMULATED MEASUREMENTS OF PHOTOGRAPHED STARS x-y IMAGE COORDINATES: MEASURED DATA FOR PROBLEM 3.3

Star Index j	Star Angular Coordinates (simulated catalog data) (Degrees)		Measured Image Coordinates* (mm)	
	α_j	δ_j	$\tilde{x}_j$	$\tilde{y}_j$
1	35	35	-9.5998	-0.8057
2	35	55	2.5214	8.4858
3	45	40	-2.7070	-2.8111
4	55	35	-8.0048	-9.7352
5	55	55	8.6610	2.2609

*These measurements were simulated as follows:

$$\left.\begin{array}{l} x_j = \\ y_j = \end{array}\right\} \text{eqns. (3.21), "perfect data"}$$

$$\left.\begin{array}{l} \tilde{x}_j = x_j + v_{xj} \\ \tilde{y}_j = y_j + v_{yj} \end{array}\right. \text{, "measured data"}$$

with

$(\phi_1, \phi_2, \phi_3) = (-35^\circ, 30^\circ, 10^\circ)$, "true camera orientation angles"

$f = 42.98$ mm = camera focal length

(v_x, v_y) = uncorrelated zero mean Gaussian random numbers

$$\sigma_{v_x} = \sigma_{v_y} = E(v_y^2)^{1/2} = E(v_x^2)^{1/2} = 2.5 \times 10^{-3} \text{ mm}$$

Example Problem 3.3

Table 3.5 summarizes the star catalog coordinates (α_j,δ_j) and image coordinate measurements $(\tilde{x}_j,\tilde{y}_j)$ for five hypothetical stars. The 43mm focal length camera implicit in the data of Table 3.5 had a true orientation (figure 3.6) of $(\phi_1,\phi_2,\phi_3) = (-35^\circ,30^\circ,10^\circ)$. Synthetic measurement errors were added to the calculated (x_j,y_j) values from eqns. (3.21) to simulate the measurement of a star photograph's image coordinates.

The Gaussian differential correction algorithm (§1.5) was applied to calculate successive approximations according to

$$\begin{Bmatrix}\phi_1\\\phi_2\\\phi_3\end{Bmatrix}^{(k+1)} = \begin{Bmatrix}\phi_1\\\phi_2\\\phi_3\end{Bmatrix}^{(k)} + \{(A^tWA)^{-1}A^tW\Delta Y\}^{(k)} \tag{3.24}$$

where

$$\Delta Y = \begin{bmatrix}\tilde{x}_1 - g_x(\alpha_1,\delta_1,\phi_1,\phi_2,\phi_3)\\ \tilde{y}_1 - g_y(\alpha_1,\delta_1,\phi_1,\phi_2,\phi_3)\\ \vdots\\ \tilde{x}_5 - g_x(\alpha_5,\delta_5,\phi_1,\phi_2,\phi_3)\\ \tilde{y}_5 - g_y(\alpha_5,\delta_5,\phi_1,\phi_2,\phi_3)\end{bmatrix} \tag{3.25}$$

$$A = \begin{bmatrix}\frac{\partial x_1}{\partial\phi_1} & \frac{\partial x_1}{\partial\phi_2} & \frac{\partial x_1}{\partial\phi_3}\\ \frac{\partial y_1}{\partial\phi_1} & \frac{\partial y_1}{\partial\phi_2} & \frac{\partial y_1}{\partial\phi_3}\\ \vdots & \vdots & \vdots\\ \frac{\partial x_5}{\partial\phi_1} & \frac{\partial x_5}{\partial\phi_2} & \frac{\partial x_5}{\partial\phi_3}\\ \frac{\partial y_5}{\partial\phi_1} & \frac{\partial y_5}{\partial\phi_2} & \frac{\partial y_5}{\partial\phi_3}\end{bmatrix} \tag{3.26}$$

$$W = \frac{1}{\sigma^2_{v_x}} \begin{bmatrix} 1_1 & & \\ & \ddots & \\ & & 1 \end{bmatrix} = (1.6 \times 10^5) \begin{bmatrix} 1_1 & & \\ & \ddots & \\ & & 1 \end{bmatrix}, \tag{3.27}$$

and the starting estimates are given as

$$\begin{Bmatrix} \phi_1 \\ \phi_2 \\ \phi_3 \end{Bmatrix}^{(0)} = \begin{Bmatrix} -25^0 \\ 20^0 \\ 7.5^0 \end{Bmatrix}. \tag{3.28}$$

Table 3.6 summarizes the convergence history of above Gaussian differential correction process. Observe that the residual variance is decreased by seven orders of magnitude after 3 corrections. The converged $(\hat{\phi}_1, \hat{\phi}_2, \hat{\phi}_3)$ compare favorably with their respective true values. Computational experience indicates that this solution is typical of the success of differential correction solutions of the *Stellar Resection Problem*. Initial orientation angle errors as large as 30^0 typically result in convergence in five or fewer corrections, so long there are two or more measured stars separated by at least $\frac{1}{2}^{\circ}$ in interstar angle.

Figure 3.8 is a star convergence trace based upon the starting, once corrected, and final converged photograph orientation angle estimates $(\hat{\phi}_1, \hat{\phi}_2, \hat{\phi}_3)$ as given in table 3.6. The third and fourth differential corrections, to plotting accuracy, overlay the measured and true images locations.

The interested reader is referred to ref. 3.4 for further discussion and more elaborate practical applications of this approach to spacecraft orientation estimation.

Example Problem 3.4: Triangulation of Orbital Photography

With reference to figure 3.9, an important problem in modern map making is the *triangulation* of photographs of the earth's surface. To appreciate the role the solution of this plays in determination of measurements, we consider the following outline

Table 3.6 DIFFERENTIAL CORRECTION OF EXAMPLE PROBLEM 3.3

	estimated angles (degrees)			star image location coordinates (measured value minus computed value in mm)										residual variance
k	$\hat{\varphi}_1$	$\hat{\varphi}_2$	$\hat{\varphi}_3$	Δx_1	Δy_1	Δx_2	Δy_2	Δx_3	Δy_3	Δx_4	Δy_4	Δx_5	Δy_5	
0	-25	20	7.5	9.67	5.45	8.28	5.62	8.99	5.68	9.20	6.26	8.57	5.73	57.67
1	-34.1286	29.5526	9.0768	.40	.60	.47	.52	.37	.54	.31	.55	.43	.44	.22
2	-34.9932	30.5267	9.9983	-.85E-2	.67E-2	-.52E-2	.88E-2	-.63E-2	.47E-2	-.68E-2	-.58E-2	-.36E-2	.59E-2	.41E-4
3	-35.0011	30.0018	10.0053	-.22E-2	-.87E-4	.27E-3	.25E-3	-.20E-3	-.18E-2	-.32E-3	-.78E-3	.25E-2	.11E-3	.21E-5
4	-35.0011	30.0018	10.0053	-.22E-2	-.87E-4	.27E-3	.25E-2	-.20E-3	-.18E-2	-.32E-3	-.78E-3	.25E-2	.11E-3	.21E-5

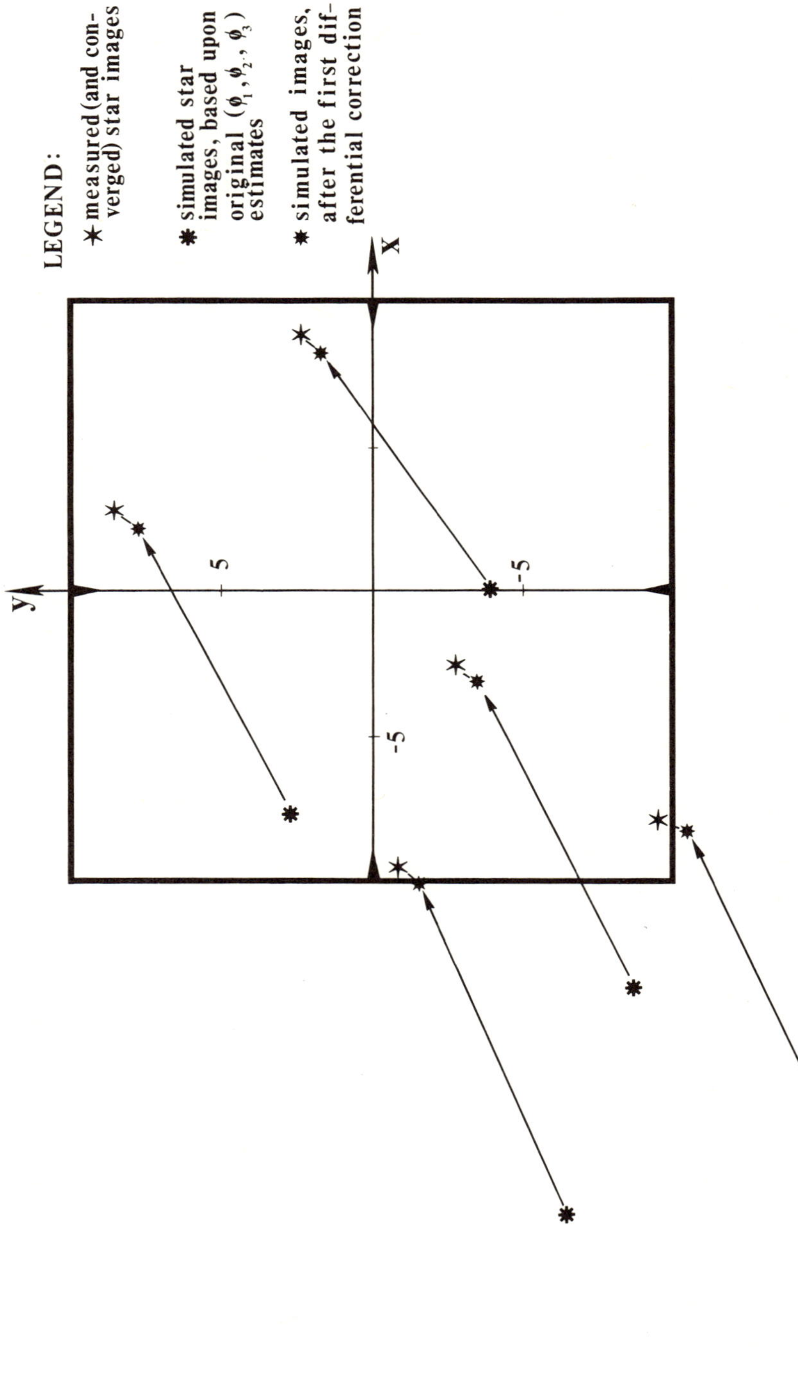

Figure 3.8 Simulated Star Images for the Least Square Differential Correction Solution of Problem 3.3

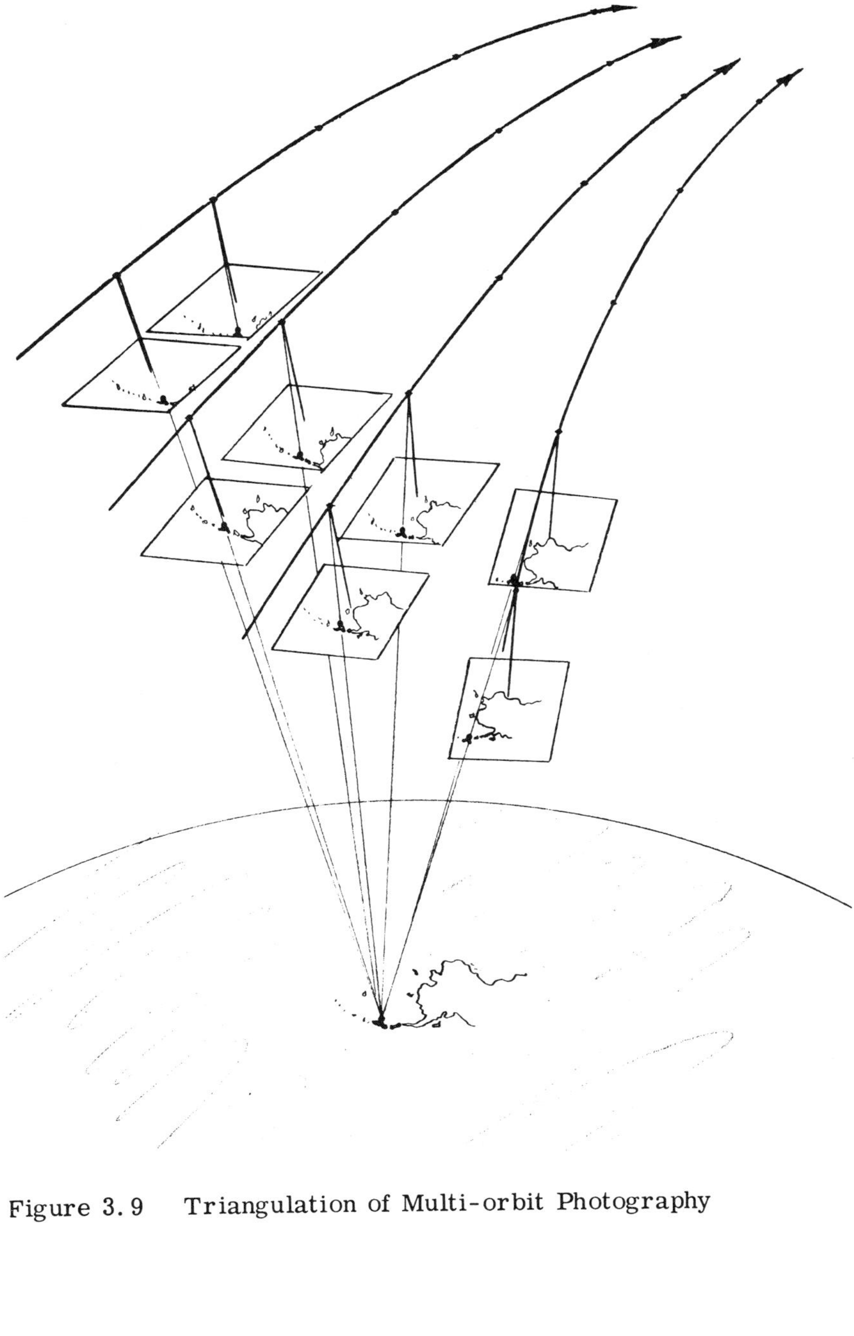

Figure 3.9 Triangulation of Multi-orbit Photography

of orbital photogrammetry/automated mapping:

- I Given photographs with overlapping fields of view, made from calibrated camera(s).
- II Determine optimal estimates of the position and orientation of the camera (usually spacecraft-fixed) at each exposure instant. This process traditionally consists of
  - A. Measure photo coordinates of identifiable ground features that have been previously surveyed *control points*.
  - B. Measure a number of photo coordinates of distinct ground features *pass points* that *have not* been previously surveyed; each pass point must be visible in at least two photographs.
  - C. *Photogrammetric Triangulation.* "Adjust" (using least square differential correction) the estimated positions and orientations of the photographs, and the coordinates of all pass points to minimize the weighted sum square of the measured-minus-computed residuals (of image coordinates).
- III Using computer automated electro-optical scanning/pattern recognition hardware, compute a dense set of ray intersection measurements of the topography.
- IV From the surface measurements of III construct a mathematical model of the surface geometry and employ this model to automate the production of topographic contour maps.

In the present section, we consider an idealized example of the photogrammetric triangulation process (II-C). In §3.5 we address the surface modeling problem IV.

The geometric optics-based projection equations central to the triangulation process can be deduced from the geometry of figures 3.10 and 3.11. This transformation is called the *terrain colinearity equations* and are derived in ref. 3.3 as

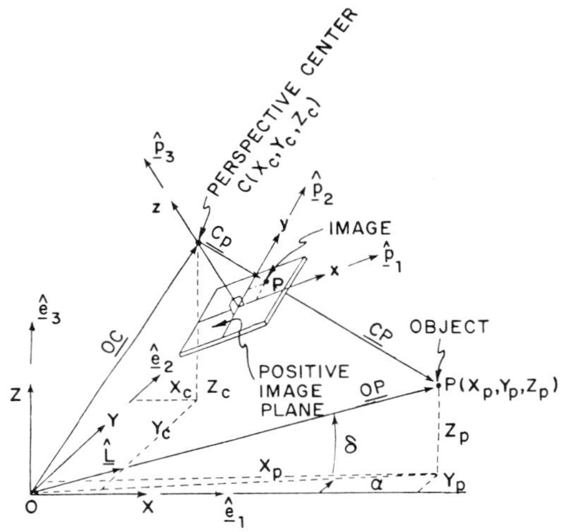

Figure 3. 10 Colinearity of the Perspective Center, Image and Terrain Feature

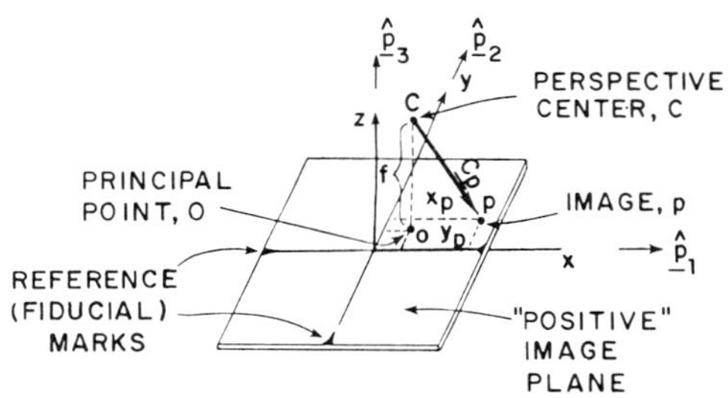

Figure 3. 11 The Photographic Coordinate system

$$x_p = x_o - f\left[\frac{C_{11}(X_p-X_c) + C_{12}(Y_p-Y_c) + C_{13}(Z_p-Z_c)}{C_{31}(X_p-X_c) + C_{32}(Y_p-Y_c) + C_{33}(Z_p-Z_c)}\right]$$

$$y_p = y_o - f\left[\frac{C_{21}(X_p-X_c) + C_{22}(Y_p-Y_c) + C_{23}(Z_p-Z_c)}{C_{31}(X_p-X_c) + C_{32}(Y_p-Y_c) + C_{33}(Z_p-Z_c)}\right]. \tag{3.29}$$

Equations (3.29) project the position of a point located at object space coordinates (X_p,Y_p,Z_p) into its image plane coordinates (x_p,y_p) for a camera with principal point offset (x_o,y_o) and focal length f. The perspective center of the camera is located at object space coordinates (X_c,Y_c,Z_c). The image space (x,y,z) axes are oriented relative to the object space (X,Y,Z) axes by the direction cosines $C_{\ell m}$; $\ell,m = 1,3$. Equations (3.29) can be written functionally as

$$\begin{aligned} x_p &= F(X_p,Y_p,Z_p,X_c,Y_c,Z_c,\phi_1,\phi_2,\phi_3,x_o,f) \\ y_p &= G(X_p,Y_p,Z_p,X_c,Y_c,Z_c,\phi_1,\phi_2,\phi_3,y_o,f). \end{aligned} \tag{3.30}$$

where ϕ_1,ϕ_2,ϕ_3 are the conventional 1-2-3 rotation sequence Euler angles parameterizing the direction cosines $C_{\ell m}$. If several points are considered in each of n photographs, eqns. (3.29) and (3.30) can be doubly subscripted to denote the equations corresponding to the ith point appearing in the jth photograph as

$$\begin{aligned} x_{p_{ij}} &= F(X_{p_i},Y_{p_i},Z_{p_i},X_{c_j},Y_{c_j},Z_{c_j},\phi_{1_j},\phi_{2_j},\phi_{3_j},x_{o_j},f_j) \\ y_{p_{ij}} &= G(X_{p_i},Y_{p_i},Z_{p_i},X_{c_i},Y_{c_j},Z_{c_j},\phi_{1_j},\phi_{2_j},\phi_{3_j},y_{o_j},f_j) \end{aligned} \tag{3.31}$$

It is noted in passing that the *Stellar colinearity equations* (3.21) can be obtained from the *terrain colinearity equations* by dividing the numerator and denominator by R (the distance $|OP|$) and letting R become large. Observe that the displacement of C from 0 becomes negligible in that $|OC| \ll R$ and note that

$$L_x = (X_p - X_c)/R$$

$$L_y = (Y_p - Y_c)/R$$

$$L_z = (Z_p - Z_c)/R$$

Equations (3.31) can be segregated into subsets depending on whether or not the particular point is a *control point* (i.e. one for which there exists apriori knowledge of its object space coordinates). Other points having distinctly measurable image plane coordinates in two or more photographs are unknown and are called *pass points*. Depending upon the number of photographs in a strip, the number of strips with sidelap, the number and distribution of control points, the number of pass points and the manner in which they are shared by adjacent photographs, an intimidating variety of observation equations of the form (3.31) can be defined in practical applications. It is not uncommon to encounter several thousand such nonlinear equations containing several hundred unknowns. An obvious computational burden is associated with the least squares solution of this class of problems.

Especially tailored least squares algorithms have been developed to solve the normal equations implicit in Gaussian differential correction of large system's of equations of the form (3.31). The partial derivative matrix (the "A matrix") is typically extremely sparse; and the *block reduction* methods of Brown (ref. 3.4 and 3.5) are a well-suited and widely-used approach to solving large least squares triangulation problems.

In references 3.5 to 3.8, methods for reducing the dimensionality of (3.31) are investigated. Reduction in dimensionality is feasible because the camera is usually spacecraft-fixed and thus must move in accord with the differential equations governing the translational/rotational motion of the satellite. Thus it is possible to "trade" the 6n unknowns

$\{X_{cj}, Y_{cj}, Z_{cj}, \phi_{1j}, \phi_{2j}, \phi_{3j}\}$, $j = 1,2,\ldots,n$ for 12 unknown initial conditions plus n exposure times. For large n, the dimensionality reduction is obviously significant. The computational implications of *dynamically constrained photogrammetry* are illustrated in another example in chapter 6. For our present discussion, let us ignore the possibility that the dimensionality may be reduced through using the spacecraft equations of motion; we consider a rather idealized moderate dimensioned example problem.

We describe here an example application in which four photographs are triangulated. The four photographs are simulations of photography from a satellite in a 100 mile circular orbit; the field of view of each photograph is approximately a 100 mile by 100 mile region on the Earth with the center of adjacent photographs being about 50 miles apart. Thus adjacent photographs have about 50% overlapping fields of view and the entire photographed region covers approximately 100 miles by 250 miles. Four *control points* (apriori surveyed points, measurable in one or more photographs) and eight *pass points* (unknown location, but measurable in two or more photographs) are visible in the photographs and located on the ground strip in such a fashion that a total of 60 measurements (30 points) result. The true camera position and orientation coordinates are given in Table 3.7, the true ground coordinates of the four control points and eight pass points are given in table 3.8. Table 3.9 contains the photograph x,y coordinates of the thirty image locations of twelve ground points (Table 3.8) as they appear in the four photographs; these values were determined from eqns. (3.29) using the data in tables (3.7) and (3.8), a 6 inch camera focal length, and a 6 inch by 6 inch photograph. The perfect data in table 3.9 were corrupted by additive Gaussian random numbers to simulate image coordinate measurement errors; the standard deviation of these errors was about 3×10^{-4} inches.

The least square differential correction algorithm (§ 1.5) is the basis of program UCPHOTO (ref. 3.8); UCPHOTO successively improves the vector of unknown parameters according to

$$\{\hat{X}\}^{(k+1)} = \{\hat{X}\}^{(k)} + \{(A^tWA)^{-1}A^tW\Delta Y\}^{(k)} \tag{3.32}$$

where

$$X^t = \{x_1\ x_2\ \ldots\ x_{12} \vdots\ x_{13}\ x_{14} \ldots\ x_{24} \vdots\ x_{25}\ x_{26}\ \ldots\ x_{48}\}$$

$$= \{\underbrace{X_{c_1}\ Y_{c_1}\ Z_{c_1} \ldots X_{c_4}\ Y_{c_4}\ Z_{c_4}}_{\text{camera position at each exposure time}} \vdots\ \underbrace{\phi_{11}\phi_{21}\phi_{31} \ldots \phi_{14}\phi_{24}\phi_{34}}_{\text{camera orientation at each exposure time}} \vdots$$

$$\underbrace{X_{p5}Y_{p5}Z_{p5} \ldots X_{p12}Y_{p12}Z_{p12}}_{\text{ground coordinates of } \textit{pass points}}\} \tag{3.33}$$

ΔY = A 60 x 1 matrix of *measured minus computed* image coordinate residuals, where the measured data are the image coordinates of table 3.9 plus the measurement errors, and the computed image coordinates are from equations (3.29).

$$A = \left[\left. \frac{\partial (x_{p_{ij}}\ y_{p_{ij}})}{\partial (x_1 \ldots . x_{48})} \right|_c \right] = \text{A 60 x 48 matrix of partial derivatives of eqns. (3.29).} \tag{3.34}$$

The A matrix structure for the current problem is rather sparse, as is illustrated in Figure 3.12. The shaded regions correspond to non-zero partial derivatives (the zeros reflect the obvious fact that only a subset of the ground points appear on any given photograph, and in the absence of dynamical constraints, the positions and orientations of the photographs are independent).

References 3.7 and 3.8 document the details of the solution of the above problem and several variations thereof. Due to the dimensionality of the problem (60 functions of 48 variables), a detailed iteration-by-iteration summary of the differential

TABLE 3.7

TRUE CAMERA EXPOSURE POSITIONS AND ORIENTATIONS FOR PROBLEM 3.4

	Photograph Numbers and Exposure Time			
	#1, $t=t_o=0$	#2, t=10.5 sec	#3, t=21 sec	#4, t=31.5 sec
X_n(miles)	0	0	0	0
Y_n(miles)	0	50.937858	101.867710	152.781552
Z_n(miles)	4063.183507	4062.864205	4061.906348	4060.310088
ϕ_1(rad)	0	-.011554	-.023116	-.034687
ϕ_2(rad)	0	.001050	.002100	.003149
ϕ_3(rad)	0	.010500	.020998	.031494

TABLE 3.8

EARTH MEASURED GROUND POINT COORDINATES FOR PROBLEM 3.4

	Ground Point Number	X_p(miles)	Y_p(miles)	Z_p(miles)
Control Points	1	-40	10	3964
	2	40	10	3963
	3	-40	140	3961
	4	40	140	3960
Pass Points	5	0	10	3963.5
	6	-40	49	3963.1
	7	0	49.5	3962.5885
	8	10	49.5	3962.4635
	9	-40	98	3961.9692
	10	0	98.5	3961.4577
	11	10	99	3961.3212
	12	0	140	3960.5

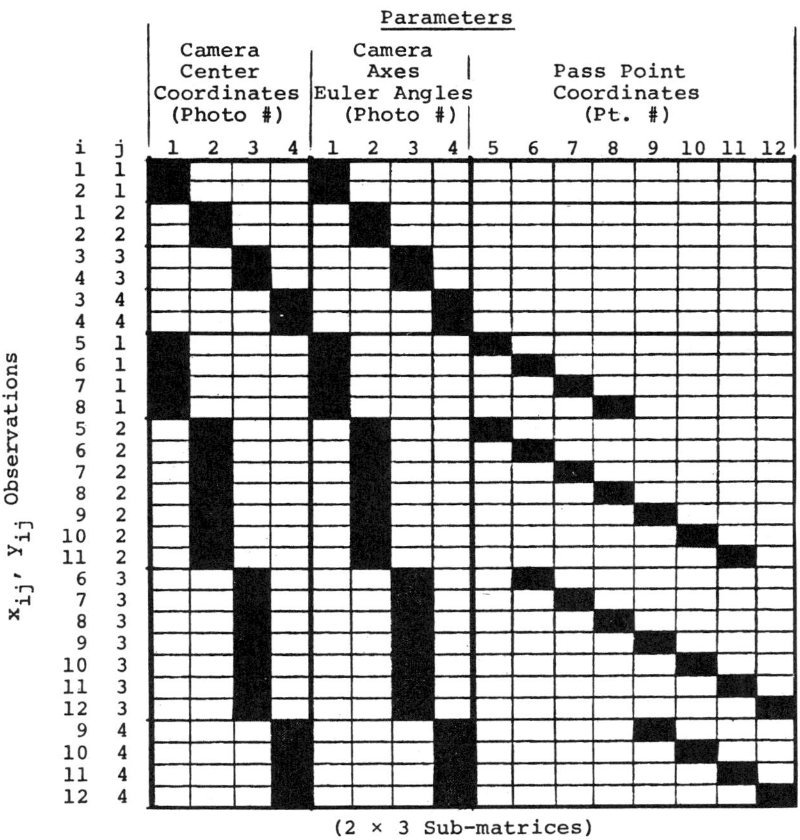

Figure 3.12 Non-zero Elements of the A Matrix for Problem 3.4

TABLE 3.9

"Perfect" Image Plane Coordinates for Problem 3.4

Ground Point Number	Photo #1 X_p(in)	Photo #1 Y_p(in)	Photo #2 X_p(in)	Photo #2 Y_p(in)	Photo #3 X_p(in)	Photo #3 Y_p(in)	Photo #4 X_p(in)	Photo #4 Y_p(in)
1	-2.419757	.604939	-2.433950	-2.377225				
2	2.395604	.598901	2.374276	-2.405076				
3					-2.333997	2.474834	-2.401944	-.985409
4					2.440882	2.355936	2.386730	-.630032
5	0	.601905	-0.18808	-2.391215				
6	-2.397998	2.937547	-2.398347	-.022009	-2.448480	-2.980837		
7	0	2.952433	.006124	-.016771	-.050153	-2.988105		
8	.595711	2.948769	.603702	-.022940	.546174	-2.997367		
9			-2.354067	2.907219	-2.386836	-.043221	-2.469292	-2.994591
10			.036740	2.898961	.011285	-.062643	-.076488	-3.027893
11			.631218	2.919072	.608188	-.045107	.519708	-3.013990
12					.063322	2.415139	.001324	-.557990

correction process is not included here. Table 3.10 summarizes the initial errors and starting errors, in the 48 element parameter vector (3.33). In a Monte Carlo test of convergence reliability, 100 different starting parameter estimates were used, the variances of the initial errors (from their respective true values) are given in Table 3.10. All one hundred starting guesses converged in five or fewer differential corrections to the same final estimates, the absolute errors (true minus converged estimates) are given in Table 3.10.

TABLE 3.10

INITIAL AND FINAL ERRORS OF 100 DIFFERENTIAL CORRECTION SOLUTIONS OF PROBLEM 3.4

Parameter Vector Subset	Initial Estimate Error Standard Deviation	Converged Estimate Error Std. Dev.
4 sets of 3 camera position coordinates	6000 ft	102 ft
4 sets of 3 camera orientation angles	3600 arc sec	47 arc sec
8 sets of 3 pass point position coordinates	6000 ft	32 ft

In other experiments the simulated image coordinate measurement error standard deviation was varied from zero to 0.1 inches. The former case resulted in the true parameters being recovered with maximum errors in the tenth significant figure (providing a basis for optimism that the colinearity equations, their partial derivatives, and the differential correction equations were correctly derived and programmed). The latter case resulted in a predictable increase (about a factor of 100) in the true-minus-converged residuals.

In chapter 6 (§6.3) the dimensionality of the current problem is reduced from 48 unknowns to 36 through the introducttion of translational/rotational dynamical constraints; without loss of accuracy. The programs used in the above discussion and complete documentation of the partial derivatives, computational tests, etc. are contained in References 3.7 and 3.8. The triangulation of aerial and orbital photography represents a field in which estimation methodology has proven extraordinarily successful; not simply in solving academic exercises like problem 3.4, but in large scale production applications. Brown's work (ref. 3.6) laid the foundation for efficient and accurate computer automated mapping of the moon, using orbital photography.

Problem 3.5 Mathematical Modeling of the Earth's Topography

A number of large irregular phenomena exist for which a satisfactory mathematical model cannot be constructed from physical laws or similar *first principles*. For example, the construction of mathematical models for the fine structure of the Earth's surface geometry does not follow from physical principles, but rather various mathematical models must be inferred from measurements of the surface geometry (i.e. survey data). In particular, triangulation/ray intersection methods using overlapping aerial or orbital photography allows collection of exceptionally dense, reasonably accurate survey measurements, as is illustrated in Figure 3.13(A).

This particular data set consists of about 60,000 measurements of about 25 square miles of topography just north of Fort Sill, Oklahoma. The relief exaggeration is about 12:1, the prominent peak is Mount Scott. Lakes Elmer Thomas and Latonka are just east and north of Mount Scott and are characterized by the data spike structure. These spurious data spikes are due to the fact that the automated pattern recognition logic fails very reliably over bland regions (such as water); the pattern recognition logic cannot reliably locate common ground features in adjacent stero photographs.

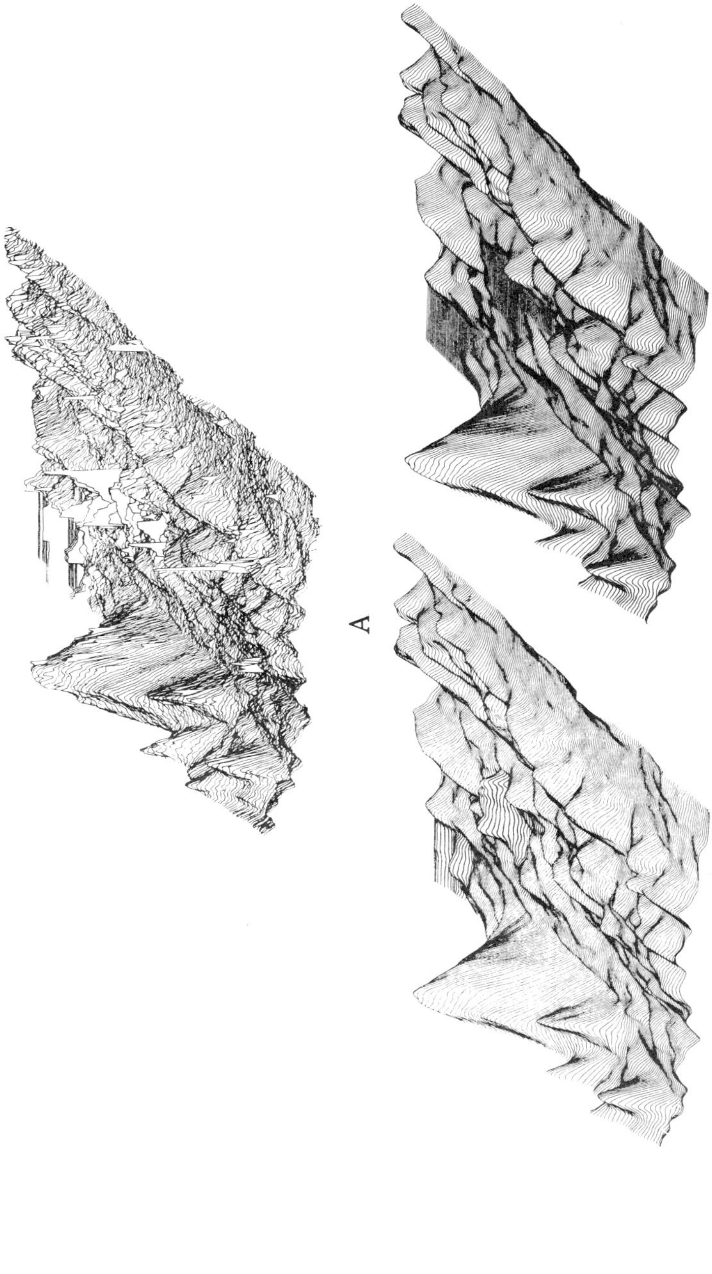

Figure 3.13 Automated Measurements of Topography via Photogrammetric Ray Intersection and therefrom Mathematical Topographic Models/Computer Automated Mapping

In the following discussion, we overview the formulation and application of a piecewise continuous approximation technique which has proven a most attractive means for modeling the Earth's topography. The formulation, computer implementation, and applications of this method are discussed in greater detail in refs. 3.9 - 3.12.

Referring to figures 3.14A and 3.14B, the central step to determining a final piecewise approximation $\bar{Z}(x,y)$, valid above the shaded unit square of figure (3.13) is to consider $\bar{Z}(x,y)$ to be a weighted average

$$\bar{Z}(x,y) = \sum_{i=1}^{4} W_i(x,y)\; Z_i(x,y) \tag{3.35}$$

of four preliminary approximations $Z_i(x,y)$; the weight functions $W_i(x,y)$ can be determined (ref. 3.9) to guarantee the following boundary conditions are satisfied by $\bar{Z}(x,y)$:

- ★ At each (say the ith) of the four corners of $\bar{Z}(x,y)$'s region of validity, $\bar{Z}(x,y)$ reduces exactly to the preliminary surface $Z_i(x,y)$, where $Z_i(x,y)$ is determined in such a fashion that its *centroid of validity* is the point (x_i, y_i).
- ★ Along each of the four boundaries of $\bar{Z}(x,y)$'s region of validity, equation (3.35) reduces exactly to

$$\bar{Z}(x,y)\Big|_{\substack{\text{along}\\ i\rightarrow j \text{ boundary}}} = W_i(x,y)Z_i(x,y) + W_j(x,y)Z_j(x,y) \tag{3.36}$$

in value and both partial derivatives.

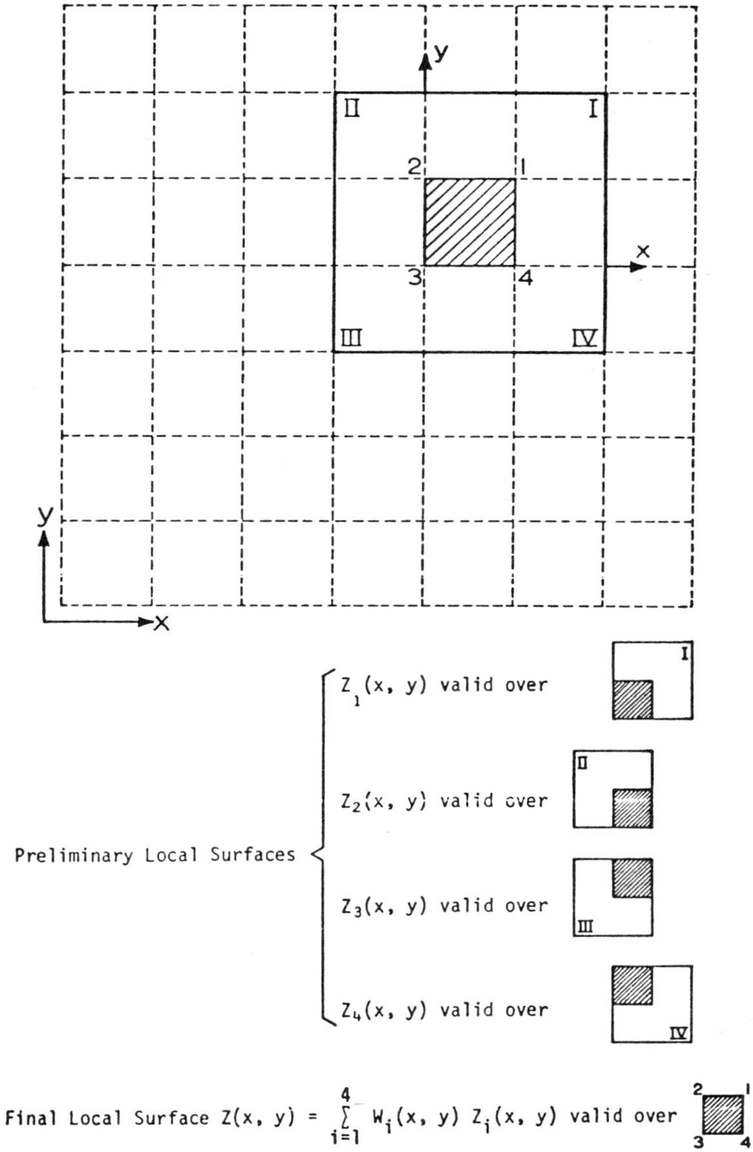

Figure 3.14A The Surface Averaging Concept

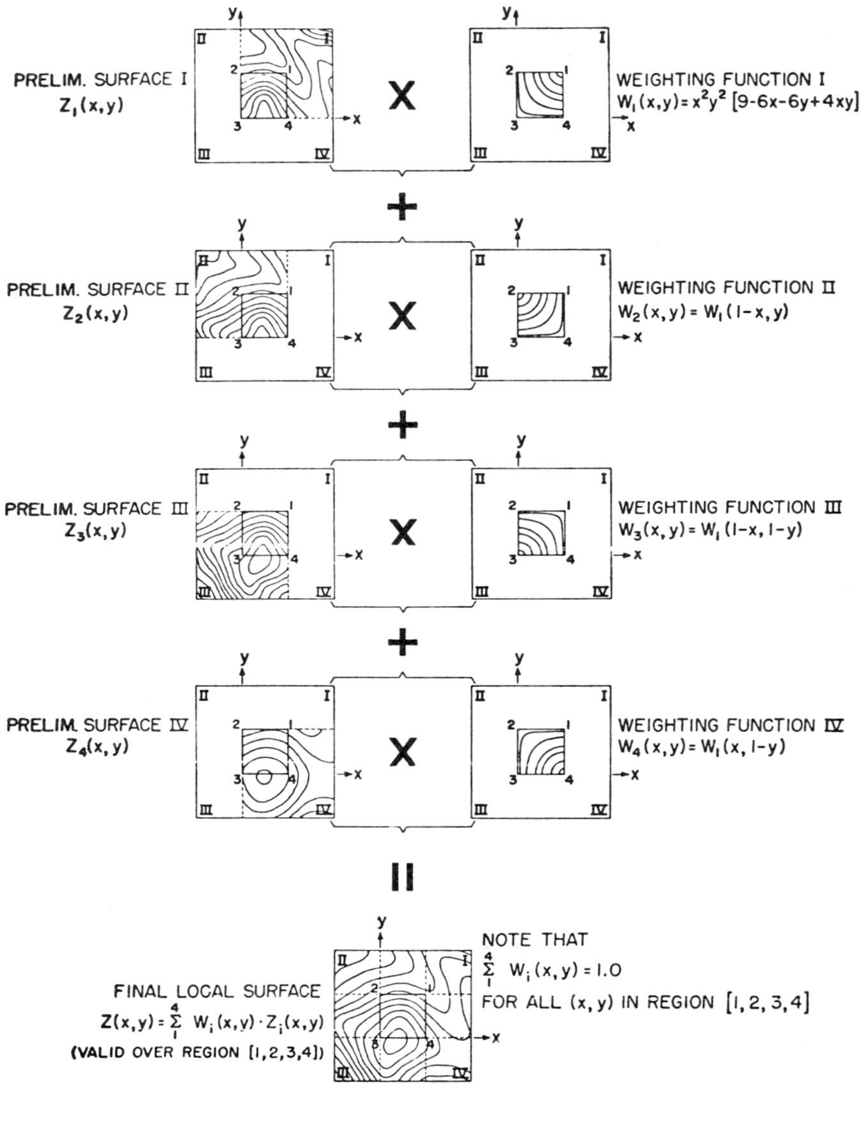

Figure 3.14B A Pictorial Representation of the Surface Averaging Process

These properties insure that adjacent local surface functions of the form (3.35) join with rigorous continuity in value and slope. The approach has been fully generalized (ref. 3.16) to allow arbitrary order derivative matching, and approximation in an arbitrary dimensioned space. A set of polynomial weight functions are determined in ref. 3.9 (for first order continuous piecewise approximation of a function of two variables) to be

$$\begin{aligned} w_1(x,y) &= x^2y^2(3-2x)(3-2y) \\ w_2(x,y) &= w_1(1-x,y) \\ w_3(x,y) &= w_1(1-x,\ 1-y) \\ w_4(x,y) &= w_1(x,\ 1-y) \quad . \end{aligned} \qquad (3.37)$$

The user has the freedom (and responsibility!) to specify the preliminary local approximations $Z_i(x,y)$ averaged in equation (3.35) in a convenient manner which is well suited to approximating the behavior of $Z(x,y)$. In particular, an attractive computational procedure results if one employs two variable polynomials

$$\begin{aligned} Z_i(x,y) &= c_{1i} + c_{2i}x + c_{3i}y + c_{4i}x^2 \\ &\quad + c_{5i}xy + c_{6i}y^2 + \ldots\ldots \quad i = 1,2,3,4 \end{aligned} \qquad (3.38)$$

then use of the polynomial weight functions (3.37) results in the final local approximations of the form (3.35) being a polynomial in x,y. This truth is most important in applications (see the root solving methods for determination of contour maps, as discussed in refs. 3.10 and 3.11).

Piecewise continuous mathematical models of the data in figure 3.12(A) were determined in references 3.10 and 3.11 using linear polynomials of the form (3.38) and the weight functions (3.37) in equation (3.35). The coefficients $\{c_{1i},\ c_{2i},\ c_{3i}\}$ of the local approximations were determined via weighted least squares as

$$\begin{Bmatrix} c_1 \\ c_2 \\ c_3 \end{Bmatrix}_i = \left[(A^t W A)^{-1} A^t W \right]_i \{\tilde{Z}\}_i \tag{3.39}$$

where

$$A_i = \begin{pmatrix} 1 & x_{1i} & y_{1i} \\ 1 & x_{2i} & y_{2i} \\ \vdots & \vdots & \vdots \\ 1 & x_{mi} & y_{mi} \end{pmatrix} \tag{3.40}$$

and

$$\{x_{1i}, y_{1i}, \tilde{z}_{1i}; \ \ldots ; \ x_{mi}, y_{mi}, \tilde{z}_{mi}\} \tag{3.41}$$

are a subset of the surface measurements near the point (x_i, y_i). For simplicity, the weight matrix may be taken as diagonal, and for the common event of equal variance (apriori estimate) measurements, then the optimum diagonal weights should be calculated from equations (3.37).

In the common event that the z measurements are taken on a grid in the x-y plane, the use of a roving local coordinate system results in the A matrix, the W matrix and therefore the matrix expression $(A^t WA)^{-1} A^t W$ being invariant for all preliminary surface fits. The practical consequences of this observation are enormous, because the coefficients of each preliminary local approximation can be efficiently calculated by simply multiplying a previously calculated matrix times the local Z-data subset. It is therefore possible to model an arbitrarily large gridded data set with one formal inversion of the normal equations.

The number of measurements approximated by a given local surface is determined by the physical size of the final region of validity. It is usually convenient to specify the region of validity size (for gridded data) by the number of data points (NFF) along the boundary of a final region of validity. Notice that the final region of validity has ¼ the area of the region spanned by each preliminary approximation.

The above weighting function technique was applied (ref. 3.11) to the measurements of 3.13A to determine piecewise continuous mathematical models. Figures 3.13B and 3.13C are profile plots of piecewise continuous models; each model contains 1050 final local functions; each final region of validity contains 64 measurements (NFF=8). The mathematical model B resulted from straight forward application of the averaging process with preliminary local surface coefficients determined as described above. Observe that the erroneous spike structure in the lakes of 3.13A resulted in erroneous "lake topography", but the representation is otherwise smooth. Excluding the lake data, the standard deviation of the data-minus-fit residuals was about 5 meters (the apriori measurement standard deviation estimate). Figure 3.13C was determined by constraining the preliminary approximations to agree with the apriori known elevation of the two lakes and their prescribed shoreline boundary.

As is evident in figure 3.15, increasing the physical size of the final region of validity serves to smooth the surface geometry approximation. For some purposes (e.g., plotting at small map scales) it may be highly desirable to suppress the fine structure meanderings of the surface. Unfortunately, such oversmoothing tends to "clip" sharp features of the topography such as mountaintops. In making aeronautical charts or military maps, this may prove catastrophic. As is evident in figure 3.16, the SAPMAP software is capable of detecting and constraining the surface model to agree with *significant* extreme features. Constraints are enforced rigorously using the constrained least square formulation of equations (1.47), (1.48), and (1.49). A rather subjective (but effective) procedure for identifying extreme features in the presence of measurement noise is discussed in ref. 3.11.

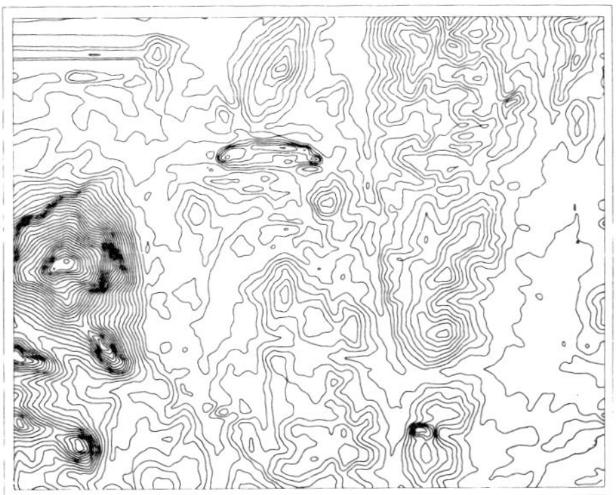

A

Contours of SAPMAP Surface Model

NFF = 6

Contour Interval 10 m

CDC 6400 CP Run Time 2.91 min.

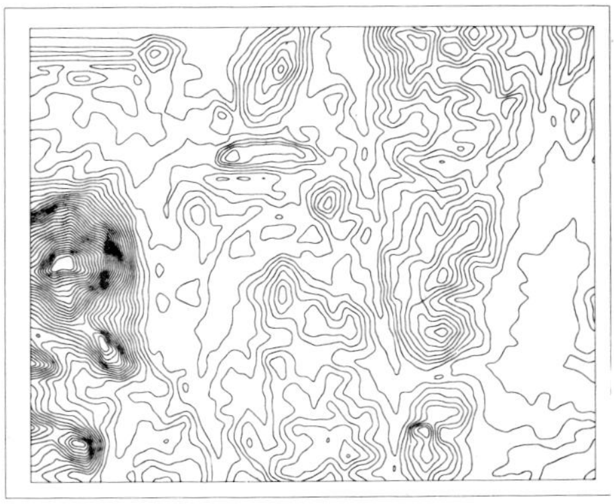

B

Contours of SAPMAP Surface Model

NFF = 8

Contour Interval 10 m

CDC 6400 CP Run Time 2.02 min.

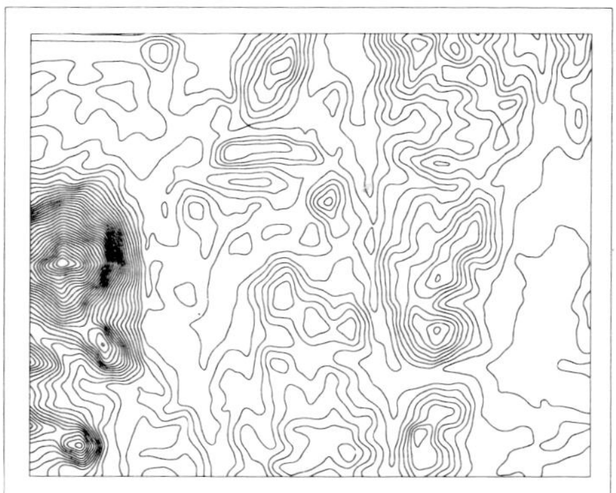

C

Contours of SAPMAP Surface Model

NFF = 10

Contour Interval 10 m

CDC 6400 CP Run Time 1.62 min.

Figure 3.15 Variable Smoothness Using the Weighting Function Technique

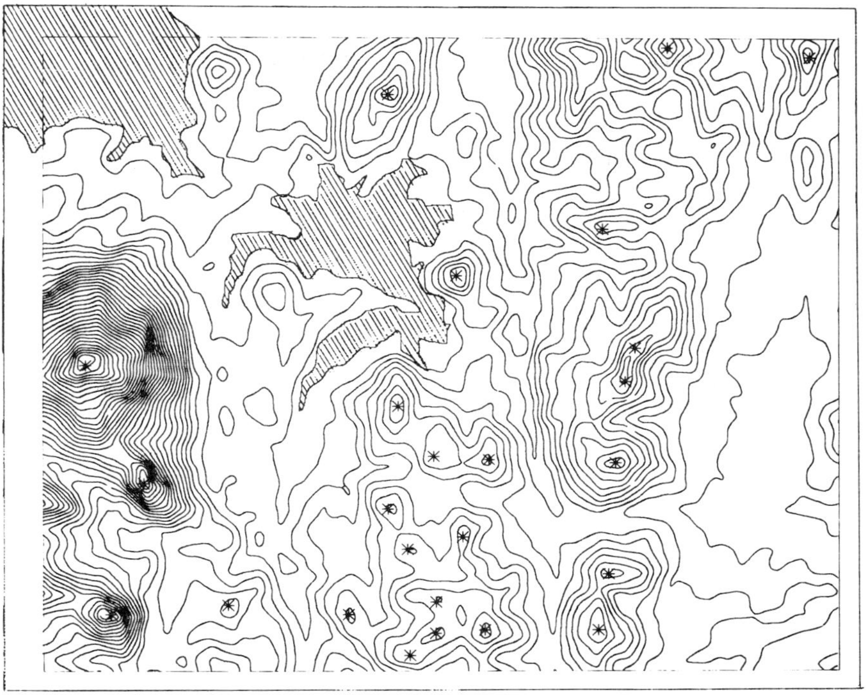

Figure 3.16 Demonstration of Peak Detection/Constraint and Lake Constraint Options of SAPMAP

NFF = 8

Contour Interval = 10 m

CDC 6400 CP Run Time 5.42 min.

3.6 Mathematical Models of the Gravitational Potential

The gravitational potential at point P near the earth is given by

$$U = G \iiint_{earth} \frac{dM}{s} = G \iiint_{earth} \frac{\eta}{s}\, dv \tag{3.42}$$

where

G = universal gravitation constant

dM = a differential earth-mass element = ηdv

dV = a differential earth-volume element

η = $\eta(x,y,z)$ = mass density of the earth at point (x,y,z).

s = distance from dM to point P.

Heiskannen and Moritz (ref. 3.13) establish that U satisfies Laplaces' differential equation

$$\nabla^2 U = \frac{1}{r^2}\frac{\partial}{\partial r}\{r^2\frac{\partial U}{\partial r}\} + \frac{1}{r\sin\theta}\frac{\partial}{\partial\theta}\{\sin\theta\frac{\partial U}{\partial\theta}\} + \frac{1}{r^2\sin\theta}\frac{\partial^2 U}{\partial\lambda^2} = 0 \tag{3.43}$$

where $\theta \equiv \pi/2 - \phi$, and establishes that the solution of (3.43) is the spherical harmonic series

$$U = \frac{\mu}{r} + \frac{\mu}{r}\sum_{n=1}^{\infty}\sum_{m=1}^{\infty}(\frac{R}{r})^n P_n^m(\sin\phi)\{C_n^m\cos m\lambda + S_n^m \sin m\lambda\} \tag{3.44}$$

where $\mu \equiv GM$ = earth's gravitational mass constant = 398,601.2 $\frac{km^3}{sec^2}$

R = reference (equatorial) earth radius = 6378.160m

$$C_n^m \equiv 2\,\frac{(n-m)!}{(n+m)!}\, G \iiint_{earth} \rho^n P_n^m(\sin\phi)\cos m\lambda\, dM$$

$$S_n^m \equiv 2\,\frac{(n-m)!}{(n+m)!}\, G \iiint_{earth} \rho^n P_n^m(\sin\phi)\sin m\lambda\, dM \tag{3.45}$$

(r,λ,ϕ) = (radial distance, longitude, lattitude)

$P_n^m(x)$ = associated Legendre functions (see ref. 3.13 and 1.2 , for detailed discussion and recursion formulas for efficient computation)

The south, east, and radial components of the local gravitational acceleration vector components follow directly from the partial differentiations:

$$G_S = -\frac{1}{r}\frac{\partial U}{\partial \phi}$$

$$G_E = \frac{1}{r\cos\phi}\frac{\partial U}{\partial \lambda} \tag{3.46}$$

$$G_R = \frac{\partial U}{\partial r}$$

Several observations are now in order:

* The south and east components of gravitational acceleration are three to four orders of magnitude smaller than the radial component, on the earth's surface, and become smaller as P moves away from the earth's surface.

* The first three significant figures of $G_R = \partial U/\partial r$ are the same all over the earth's surface and are given by the *point mass potential*

  $$\frac{\partial}{\partial r}\left(\frac{GM}{r}\right)\bigg|_{r=R} = -\frac{GM}{R^2} \cong -9.80 \times 10^{-3} \text{ m/sec}^2$$

  which results in ignoring all but the first term of the spherical harmonic series (3.44).

* The spherical harmonic coefficients (3.45) cannot be evaluated formally by carrying out the integrals, because the earth's mass distribution is not known with sufficient accuracy. These coefficients must be estimated indirectly, by truncating the series (3.44) and iteratively improving (via least squares differential correction) the coefficients of the truncated series to make numerically integrated satellite motion

agree in the least square sense with observed satellite motion. Modern gravity field models have several hundred non-zero terms. As was anticipated in the first example of chapter 1, truncating mathematical models causes an uncertain degree of aliasing of the estimated parameters. Despite these difficulties, independently determined gravity fields are usually in agreement to seven or more significant figures above 100 miles from the earth's surface; even though the higher order coefficients are aliased and not always in close agreement.

The demand for a seven plus significant figure gravity model is generated by the necessity for highly accurate satellite and missile trajectory integration. The above series (3.43) and several variations thereof (see references 3.14 and 3.15) serve as the most commonly used gravity model for satellite orbit integrations. While *satisfactory*, this approach seems far from optimal.

In spite of the success achieved in developing feasible algorithms based upon (3.44) and (3.46) and analogous series, the increasing burden of computing acceleration from ever more lengthly global gravity models consumes an even larger fraction of the central processor time in trajectory/orbit integrations. This fact and other considerations has motivated research (ref. 3.15, 3.16) into other possible global gravity representations.

The present discussion departs from current practice by separating from the onset the two important questions:

(1) What gravity representation should be fit to observed data to estimate unknown model constants from satellite observations and/or surface gravimetry (and thereby establish a global gravity field approximation)?

(2) In highly repetitive gravitational calculations (e.g. computing acceleration for use in integration of satellite orbits/missile trajectories), what gravity model is most efficient computationally?

In gravitational modeling research to date, both questions have been (perhaps subconsciously) addressed simultaneously. Explicit separation of the "optimum determination" and "optimum use" questions appears to be a most important consideration in further refinement of existing gravity models to accommodate the ever more precise and abundant observed data. Without engaging in the important quest for a model which can be best determined from observed data, we address the problem of determining an "optimum use" gravitational model.

THE FINITE ELEMENT APPROACH

The dominate global features of the gravity field are efficiently represented by the dominant low degree terms in equation (3.44). This fact motivates the segregation of the total gravitational potential at a point (r, ϕ, λ) as

$$U = U_{REF}\,(r, \phi, \lambda) + \Delta U(r, \phi, \lambda) \tag{3.48}$$

where it is understood that a single low degree truncation of equation (3.44) is adopted to define U_{REF} globally, but instead of attempting to model "everything else" as a single global series, we shall determine a global family of locally valid disturbance functions to model ΔU and its gradient. Since the local equations must model only the gravity undulations (in addition to U_{REF}) in a specific local volume, it is reasonable to anticipate significantly more compact expressions than a single global expansion of comparable local accuracy. This is, of course, the thesis of the finite element approach.

Having decided to pursue the finite element approach, it is necessary to make several important decisions; it is necessary to define

* What portion of the geopotential is to be approximated as ΔU in equation (3.48).
* The specific mathematical structure to use as local approximations of ΔU and its gradient.
* The size of the finite elements.
* The procedure to be used in determining numerical values for coefficients in the local approximations.
* The order of continuity requirements between adjacent approximations across their mutual boundaries of validity.
* Procedures to evaluate the validity of the finite element model.

These decisions are coupled and effect in a complicated way the accuracy and efficiency of the resulting finite element model. Attempts to analytically resolve these issues apriori proved unsuccessful. Guided by intuition, we have developed systematically collected empirical data, leading to the development of a prototype finite element model of the geopotential. We overview these results, considering several approaches for constructing gravitational finite element models and summarize numerical experiments with them.

A WEIGHTING FUNCTION APPROACH TO PIECEWISE CONTINUOUS APPROXIMATION

A recent paper (3.16) documents the theoretical development and application of a versatile piecewise continuous approximation technique. This *weighting function approach* is a 3 variable extension of the method presented in §3.5; it determines an arbitrarily large family of locally valid functions which join with rigorous piecewise continuity through any prescribed order of partial differentiation. As applied to the approximation of a three variable function, $F(X_1, X_2, X_3)$ each *final local approxi-*

mation $\bar{F}(X_1, X_2, X_3)$ is defined as a weighted average of eight *preliminary local approximations* $\{f_{ijk}(X_1, X_2, X_3);\ i,j,k = 0, 1\}$ as

$$\bar{F}(X_1, X_2, X_3) = \sum_{i=o}^{1} \sum_{j=o}^{1} \sum_{k=o}^{1} W_{ijk}(X_1, X_2, X_3) \cdot f_{ijk}(X_1,X_2,X_3) \tag{3.49}$$

The eight preliminary approximations may have arbitrary mathematical structure (a most flexible and powerful feature of this approach) but must be determined in such a fashion that their respective *centroids of validity* lie at the eight vertices of a parallelopiped (in X_1, X_2, X_3 space), which defines the volume in which the final local approximation $\bar{F}(X_1, X_2, X_3)$ is valid. The form of the weight functions can be selected (ref. 3.16) to guarantee that $\bar{F}$ and its similarily determined adjacent approximations are continuous across their mutual boundaries of validity; the order of partial differentiation to which continuity is desired is controlled by selecting appropriate weight functions. Specifically, table 3.11 gives the weight function $W_{111}(X_1, X_2, X_3)$ for various orders of continuity*. The remaining seven weight functions are obtained by reflecting $W_{111}(X_1, X_2, X_3)$ as

$$\begin{aligned}
W_{000}(X_1, X_2, X_3) &= W_{111}(1-X_1, 1-X_2, 1-X_3)\\
W_{001}(X_1, X_2, X_3) &= W_{111}(1-X_1, 1-X_2, X_3)\\
W_{010}(X_1, X_2, X_3) &= W_{111}(1-X_1, X_2, 1-X_3)\\
W_{100}(X_1, X_2, X_3) &= W_{111}(X_1, 1-X_2, 1-X_3)\\
W_{011}(X_1, X_2, X_3) &= W_{111}(1-X_1, X_2, X_3)\\
W_{101}(X_1, X_2, X_3) &= W_{111}(X_1, 1-X_2, X_3)\\
W_{110}(X_1, X_2, X_3) &= W_{111}(X_1, X_2, 1-X_3)
\end{aligned} \tag{3.50}$$

---

*Under the assumption that X_1, X_2, X_3 are non-dimensional local coordinates with ($X_1 = i$, $X_2 = j$, $X_3 = k$) locating the eight vertices of the unit *cube of validity* of $\bar{F}(X_1, X_2, X_3)$.

The weight functions are positive and satisfy the constraint that

$$\sum_{i=o}^{1} \sum_{j=o}^{1} \sum_{k=o}^{1} W_{ijk} (X_1, X_2, X_3) = 1 \tag{3.51}$$

They may be interpreted geometrically as follows: The maximum value W_{ijk} is unity which occurs at $X_1 = i$, $X_2 = j$, $X_3 = k$; the surfaces of constant weight are spherical in the vicinity of (i,j,k) but become increasingly angular until the surface of zero weight is the walls of the cube opposite to i,j,k. Thus, in equation (3.49), $\underline{W}_{ijk}$ causes the preliminary local approximation f_{ijk} to dominate $\bar{F}$ in the vicinity of f_{ijk}'s centroid of validity, but has no effect on $\bar{F}$ (in value or first m partial derivatives) along the opposite cell boundary. The key to piecewise continuity is the fact that $\bar{F}$ is completely defined (on each of the six boundary "walls") by the four preliminary approximations whose centroids are the vertices of the respective walls. Since the four preliminary approximations (defining any given cell wall) are shared by adjacent final approximations, it is clear that piecewise continuity is assured.

LOCAL GRAVITY APPROXIMATIONS VIA TAYLOR'S SERIES

Given an apriori-determined global gravity model (e.g., a spherical harmonic series), perhaps the most obvious strategy of generating local approximations is by Taylor's series. The disturbance potential and its gradient can be locally approximated as the truncated Taylor's series of three variables

$$\underline{\Delta G}_{ijk}(r,\phi,\lambda) \cong \underline{\Delta G}(r_i,\phi_j,\lambda_k) + \sum_{n=1}^{M} \sum_{I=1}^{n} \sum_{J=1}^{n-I} \underline{G}_{IJK} X_1^I X_2^J X_3^K \tag{3.52}$$

where

$$\underline{\Delta G}_{ijk} \equiv \begin{Bmatrix} \Delta U \\ -\dfrac{1}{r} \dfrac{\partial(\Delta U)}{\partial\phi} \\ \dfrac{1}{r\cos\phi} \dfrac{\partial(\Delta U)}{\partial\lambda} \\ \dfrac{\partial(\Delta U)}{\partial r} \end{Bmatrix} \equiv \begin{Bmatrix} \Delta U \\ \Delta G_S \\ \Delta G_E \\ \Delta G_R \end{Bmatrix} \tag{3.53}$$

$$\underline{G}_{IJK} \quad \frac{(\Delta r)^I (\Delta\phi)^J (\Delta\lambda)^K}{I! \quad J! \quad K!} \quad \frac{\partial^n}{\partial r^I \partial\phi^J \partial\lambda^K} \begin{Bmatrix} \Delta U \\ \Delta G_S \\ \Delta G_E \\ \Delta G_R \end{Bmatrix}_{(r,\phi,\lambda)=(r_i,\phi_j,\lambda_k)} \tag{3.54}$$

$K = n-I-J$, M = order of the local Taylor's series.

(r_i,ϕ_j,λ_k) = an arbitrary local expansion point {which is the *centroid of validity* of (3.52)}

$(2\Delta r,\ 2\Delta\phi,\ 2\Delta\lambda)$ = dimensions of the region of validity of (3.52), and

$(X_1,\ X_2,\ X_3) \equiv (\dfrac{r-r_i}{\Delta r},\ \dfrac{\phi-\phi_j}{\Delta\phi},\ \dfrac{\lambda-\lambda_k}{\Delta\lambda})$ = non-dimensional local coordinates.

The partial derivatives (3.54) can be rigorously derived from the parent global model of ΔU. Reference 3.18 gives analytical expressions for computation of the partial derivatives specifically for the case that the parent global model of ΔU is a spherical harmonic series. It should be pointed out that specific numerical values for the elements of $\underline{G}_{IJK}$ are computed apriori and stored; for centroids of validity distributed over the (r,ϕ,λ) space according to some specified pattern.

In using (say) the foregoing weighting function formulation to compute local disturbance acceleration (from a piecewise continuous, finite element model), the appropriate, previously computed, set of $\underline{G}_{IJK}$ coefficients are employed to compute equations (3.52) as preliminary local approximations for substitution into

$$\begin{Bmatrix} U(r,\phi,\lambda) \\ G_S(r,\phi,\lambda) \\ G_E(r,\phi,\lambda) \\ G_R(r,\phi,\lambda) \end{Bmatrix} = \begin{Bmatrix} U_{REF}(r,\phi,\lambda) \\ \frac{1}{r}\frac{\partial U_{REF}}{\partial \phi} \\ \frac{1}{r\cos\phi}\frac{\partial U_{REF}}{\partial \lambda} \\ \frac{\partial U_{REF}}{\partial r} \end{Bmatrix}$$

$$+ \sum_{i=o}^{1} \sum_{j=o}^{1} \sum_{k=o}^{1} W_{ijk}(X_1,X_2,X_3)\, \underline{\Delta G}_{ijk}(r,\phi,\lambda) \qquad (3.55)$$

where

$X_1 = (r-r_o)/\Delta r$

$X_2 = (\phi-\phi_o)/\Delta\phi$ Non-dimensional local coordinates

$X_3 = (\lambda-\lambda_o)/\Delta\lambda$

(r_o,ϕ_o,λ_o) coordinates of the "lower left corner" of equation (3.55)'s region of validity: $\{0 \le X_i \le 1;\quad i = 1,2,3\}$

The gravity representation (3.55) leads to a non-uniform distribution of errors, observe that the approximations become exact as the displacement of the evaluation point from a centroid of validity (expansion point) decreases to zero; but more generally, the final approximation is the average of eight approximations containing errors. This observation led us to expect from the onset that Taylor's series would not be the optimum choice of preliminary approximation functions; one should seek preliminary approximations with more uniform error distributions.

LOCAL GRAVITY APPROXIMATIONS VIA LEAST SQUARES APPROXIMATION

As an alternative to the local Taylor's series {equation (3.52) we consider as the local model of disturbance gravity

$$\begin{Bmatrix} \Delta U \\ \Delta G_S \\ \Delta G_E \\ \Delta G_R \end{Bmatrix} = \sum_{n=o}^{M} \sum_{I=o}^{n} \sum_{J=o}^{n-I} \begin{Bmatrix} U_{IJK} \\ S_{IJK} \\ E_{IJK} \\ R_{IJK} \end{Bmatrix} F_{IJK}(X_1, X_2, X_3) \tag{3.56}$$

where

$K = n-I-J$

$\{F_{000}; F_{001}, F_{010}, F_{100}; F_{002}, F_{011}, F_{101}, F_{020}, F_{110}, F_{200}; \ldots, F_{M00}(X_1, X_2, X_3)\}$

are a suitable set of linearly independent *basis functions*,

and

$$\{U\}^T = \{U_{000}\ U_{001}\ U_{010}\ U_{100}\ \cdots\ U_{M00}\},$$

$$\{S\}^T = \{S_{000}\ S_{001}\ S_{010}\ S_{100}\ \cdots\ S_{M00}\},$$

$$\{E\}^T = \{E_{000}\ E_{001}\ E_{010}\ E_{100}\ \cdots\ E_{M00}\},$$

$$\{R\}^T = \{R_{000}\ R_{001}\ R_{010}\ R_{100}\ \cdots\ R_{M00}\},$$

are coefficients determined so that the sum square residual error {between equation (3.56) and the parent model of disturbance gravity} is minimized.

In particular, if the least square coefficient estimates are determined by fitting equation (3.56) to local evaluations of a global gravity model, then the coefficient estimates are given by the normal equations (§1.2)

$$\begin{aligned} \{U\} &= \{B\}\ \{\Delta U_c\} \\ \{S\} &= \{B\}\ \{\Delta GS_c\} \\ \{E\} &= \{B\}\ \{\Delta GE_c\} \\ \{R\} &= \{B\}\ \{\Delta GR_c\} \end{aligned} \tag{3.57}$$

where

$$\{B\} \equiv \{(A^T WA)^{-1}\ A^T\ W\} \tag{3.58}$$

TABLE 3.11

WEIGHT FUNCTION W_{111} FOR VARIOUS ORDERS OF CONTINUITY

Order of Continuity m	$W_{111}(X_1, X_2, X_3)$
0	$X_1\ X_2\ X_3$
1	$\{X_1^2(3-2X_1)\}\{X_2^2(3-2X_2)\}\{X_3^2(3-2X_3)\}$
2	$\{X_1^3(10-15X_1 + 6X_1^2)\}\{X_2^3(10-15X_2 + 6X_2^2)\}\{X_3^2(10-15X_3 + 6X_2^2)\}$
3	$\pi_{i=1}^{3} \{X_i^4(35-84X_i + 70X_i^2 - 20X_i^3)\}$
.	.
.	.
.	.
m	$\pi_{i=1}^{3} \left\{ \frac{(2m+1)!(-1)^m}{(m!)^2} \sum_{r=o}^{m} \frac{(-1)^r \binom{m}{r}}{2m - r + 1} X_i^{2m - r + 1} \right\}$

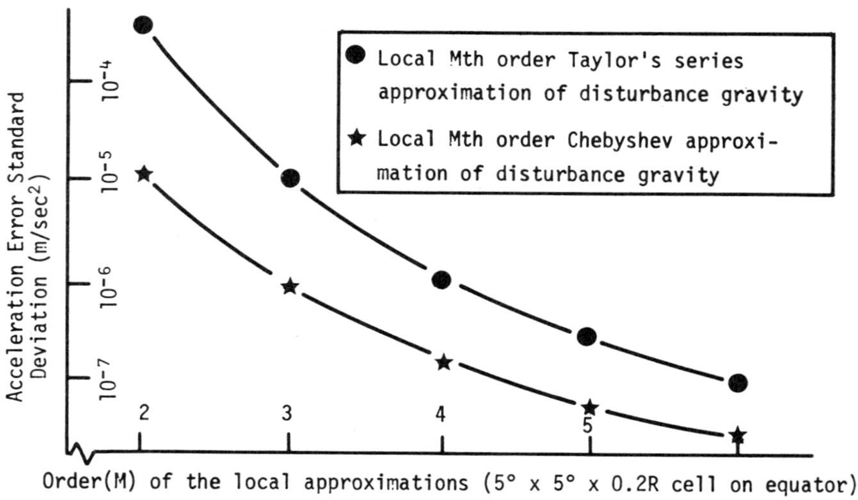

Figure 3.17 Chebyshev Polynomials vs Taylor's Series Gravity Approximations

TABLE 3.12

CHEBYSHEV POLYNOMIALS

n	$t_n(x)$	$\frac{dt_n}{dx}$
0	1	0
1	x	1
2	$2x^2 - 1$	$4x$
3	$4x^3 - 3x$	$12x^2 - 3$
.	.	.
.	.	.
.	.	.

RECURSIONS:

$$t_n(x) = 2x\, t_{n-1}(x) - t_{n-2}(x) \qquad -1 \le x \le 1$$

$$\frac{dt_n}{dx} = 2t_{n-1}(x) + 2x \frac{dt_{n-1}}{dx} - \frac{dt_{n-2}}{dx}$$

SHIFTED CHEBYSHEV POLYNOMIALS:

$$T_n(y) = t_n(2y - 1) \qquad 0 \le y \le 1$$

$$\{A\} \equiv \begin{bmatrix} F_{000}(X_{11},X_{21},X_{31}) & F_{100}(X_{11},X_{21},X_{31}) & \cdots & F_{M00}(X_{11},X_{21},X_{31}) \\ F_{000}(X_{12},X_{22},X_{32}) & F_{100}(X_{12},X_{22},X_{32}) & \cdots & F_{M00}(X_{12},X_{22},X_{32}) \\ \vdots & \vdots & \ddots & \vdots \\ F_{000}(X_{1n},X_{2n},X_{3n}) & F_{100}(X_{1n},X_{2n},X_{3n}) & \cdots & F_{M00}(X_{1n},X_{2n},X_{3n}) \end{bmatrix} \tag{3.59}$$

$\{W\} = \{w_{ij}\}$ = an n x n positive definite weight matrix

$$\begin{aligned} \{\Delta U_c\}^T &\equiv \{\Delta U(X_{11},X_{21},X_{31})\ \Delta U(X_{12},X_{22},X_{32})\ \cdots\ \Delta U(X_{1n},X_{2n},X_{3n})\} \\ \{\Delta GS_c\}^T &\equiv \{\Delta G_S(X_{11},X_{21},X_{31})\ \Delta G_S(X_{12},X_{22},X_{32})\ \cdots\ \Delta G_S(X_{1n},X_{2n},X_{3n})\} \\ \{\Delta GE_c\}^T &\equiv \{\Delta G_E(X_{11},X_{21},X_{31})\ \Delta G_E(X_{12},X_{22},X_{32})\ \cdots\ \Delta G_E(X_{1n},X_{2n},X_{3n})\} \\ \{\Delta GR_c\}^T &\equiv \{\Delta G_R(X_{11},X_{21},X_{31})\ \Delta G_R(X_{12},X_{22},X_{32})\ \cdots\ \Delta G_R(X_{1n},X_{2n},X_{3n})\} \end{aligned} \tag{3.60}$$

are local evaluations of disturbance potential and acceleration (from a parent, global gravity model) at the set of points

$$(X_{1\ell}, X_{2\ell}, X_{3\ell}) = \left(\frac{r_\ell - r_i}{\Delta r}, \frac{\phi_\ell - \phi_j}{\Delta\phi}, \frac{\lambda_\ell - \lambda_k}{\Delta\lambda}\right) ; \ \ell = 1,2,\ldots,n \tag{3.61}$$

in the vicinity of the centroid of validity (r_i, ϕ_j, λ_k). It has been found advantageous to generate the data (3.60) on a uniform grid in the local $(X_{1\ell}, X_{2\ell}, X_{3\ell})$ space; providing this grid is held identical for all cells, then equations (3.59) and (3.58) can be computed <u>only once</u> and simply reused in (3.57) to operate upon appropriate local data to generate the entire global set of local coefficients for use in equation (3.56).

An infinity of choices exist for the basis functions in (3.56); after some experimentation, we selected the set of all Chebyshev polynomial products up to M^{th} order (Table 3.12) as

$$\begin{Bmatrix} F_{000} \\ F_{001} \\ F_{010} \\ F_{100} \\ F_{002} \\ F_{011} \\ \cdot \\ \cdot \\ \cdot \\ F_{IJK} \\ \cdot \\ \cdot \\ \cdot \\ F_{M00} \end{Bmatrix} = \begin{Bmatrix} 1 \\ T_1(X_3) \\ T_1(X_2) \\ T_1(X_1) \\ T_2(X_3) \\ T_1(X_2)\ T_1(X_3) \\ \cdot \\ \cdot \\ \cdot \\ T_I\ (X_1)\ T_J\ (X_2)\ T_K\ (X_3) \\ \cdot \\ \cdot \\ \cdot \\ T_M(X_1) \end{Bmatrix} \qquad (3.62)$$

NUMERICAL TRADEOFF STUDIES

To gather the empirical evidence upon which to base selections from the many alternatives implicit in the above developments, a versatile computer software system has been developed. Figure 3.17 summarizes a portion of a tradeoff study whose objective is to help decide whether Taylor's series or locally fit Chebyshev polynomials should be employed as local gravity approximations. The RMS of the acceleration error norm was determined by computing the variance of a directly evaluated acceleration error sample. In this case, the sample consisted of 216 uniformly spaced error calculation points over each finite element. Using the RMS acceleration error norm as an accuracy criterion (for a given order of the local approximation) it is clear that the locally fit Chebyshev polynomials are superior to equal order Taylor's series. The very non-uniform error distribution characteristic of truncated Taylor's series (zero error at expansion point, but rapidly degrading away from that point) is partially compensated for in the weighting function approach by the bell shaped weights and redundancy of the method. However, the RMS of the locally fit Chebyshev polynomials have been found consistently superior for low order approximations (≤ 4), usually by an order of magnitude. Consideration of other goodness of fit criteria (smallness of mean error, smallness of maximum error, near Gaussian residuals, etc.), support the choice of locally fit polynomials over local Taylor's series.

Based upon the numerical experiments done this far, the following conclusions were drawn (ref. 3.17):

(1) Rigorous piecewise continuity of the local approximations, while desirable from conceptual and esthetic viewpoints, appears weakly justified using the small RMS error criterion. The weighted average of eight local approximations is consistently superior to any of the original approximations, but often not sufficiently superior to justify the 8:1 redundancy of the method.

(2) By fixing the order of the local approximations and selecting a specific accuracy criterion, straightforward variation in the finite element dimensions leads quickly to the family of maximum volume elements consistent with these two constraints. Global numerical tests support the conclusion that global decisions can usually be reliably made based upon local numerical experiments, so long as the local experiments include a full range of latitude variation.

(3) Analysis of repeated statistical samples of acceleration errors, (between local Chebyshev approximations and a degree 23 global spherical harmonic series) taken from various finite elements support the following conclusions regarding distribution of approximation errors:

(3a) The mean acceleration error has *always* been found to be at least one order of magnitude less than the sample RMS acceleration error, providing the sample was taken from at least 200 sample points located either on a uniform grid or randomly located within the element.

(3b) The maximum acceleration error has *always* been found to be less than 4 times the sample RMS acceleration error, with the same sample restrictions as (3a).

A PROTOTYPE FINITE ELEMENT MODEL OF THE GEOPOTENTIAL

Based upon insight gained in parametric studies with the experimental software, a prototype finite element model has been developed (ref. 3.17). The global finite model has the following structure:

$$\begin{Bmatrix} U(r,\phi,\lambda) \\ G_S(r,\phi,\lambda) \\ G_E(r,\phi,\lambda) \\ G_R(r,\phi,\lambda) \end{Bmatrix} = \begin{Bmatrix} U(r,\phi,\lambda) \\ G_S(r,\phi,\lambda) \\ G_E(r,\phi,\lambda) \\ G_R(r,\phi,\lambda) \end{Bmatrix}_{REF} + \sum_{n=o}^{M} \sum_{I=o}^{n} \sum_{J=o}^{n-I} \begin{Bmatrix} U_{IJK} \\ S_{IJK} \\ E_{IJK} \\ R_{IJK} \end{Bmatrix} T_I(X_1)T_J(X_2)T_K(X_3) \tag{3.63}$$

where

$$U_{REF}(r,\phi,\lambda) \equiv \frac{GM}{r} \{1 + C_2^o \left(\frac{R}{r}\right)^2 P_2^o(\sin\phi)\} \tag{3.64}$$

$$G_{S_{REF}} = -\frac{1}{r}\frac{\partial U_{REF}}{\partial \phi} = \frac{-GM}{r^2} C_2^o \left(\frac{R}{r}\right)^2 \frac{d}{d\phi} \{P_2^o(\sin\phi)\} \tag{3.65}$$

$$G_{E_{REF}} = \frac{1}{r\cos\phi} \frac{\partial U_{REF}}{\partial \lambda} = 0 \tag{3.66}$$

$$G_{R_{REF}} = \frac{\partial U_{REF}}{\partial r} = -\frac{GM}{r^2}\{1 + 3\ C_2^o \left(\frac{R}{r}\right)^2 P_2^o(\sin\phi)\} \tag{3.67}$$

and U_{IJK}, S_{IJK}, E_{IJK}, R_{IJK} are the appropriate set of apriori computed local coefficients; computed according to equations (3.59), (3.58), and (3.57) with

$$F_{IJK}(X_1, X_2, X_3) = T_I(X_1)\ T_J(X_2)\ T_K(X_3) \tag{3.68}$$

to accurately replace the spherical harmonic model (ref. 3.17) of disturbance gravity

$$U = \frac{GM}{r} \sum_{n=2}^{23} \sum_{m=1}^{n} \left(\frac{R}{r}\right)^n P_n^m(\sin\phi)\ \{C_n^m \cos m\lambda + S_n^m \sin m\lambda\} \tag{3.69}$$

and its gradient.

Figure 3.18 is a projection of a global contour map of the radial disturbance acceleration $\{\frac{\partial(\Delta U)}{\partial r}\}$ on the earth's surface.

To fully define the procedure for constructing the finite elements, the following decisions were made:

(1) Develop a finite element model for the spherical shell from 1 to 1.2 earth radii.

(2) Adjust the finite element size and/or the order of the local approximation so that the acceleration approximation errors enter (at worst) in the seventh significant figure.

(3) Fix the order of the local approximations at $M = 3$ {in equation (3.64)}.

Holding the radial dimension fixed at 0.2 R and adjusting the longitude by lattitude dimensions to maintain requirement (2) led to the set of 1500 finite elements whose bases are shown in the flat projection of Figure 3.19.

The acceleration error residual norm

$$\Delta G = \{(GSH\text{-}GFE)^T\ (GSH\text{-}GFE)\}^{\frac{1}{2}} \tag{3.70}$$

where

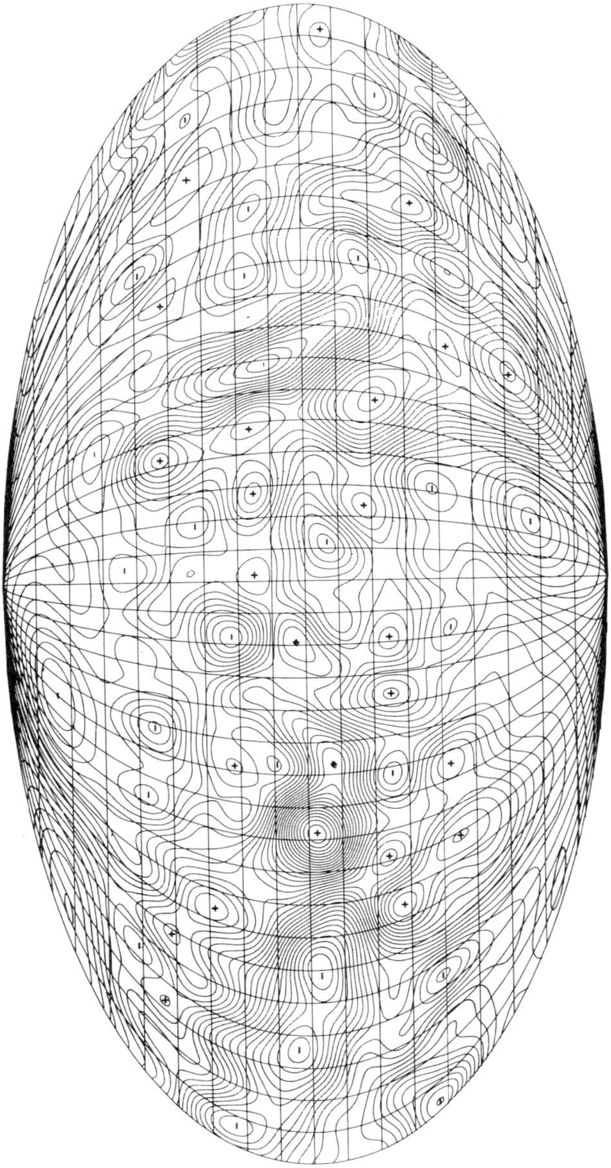

Figure 3.18 Radial Disturbance Acceleration $\frac{\partial}{\partial r}(\Delta U)$ on the Earth's Surface (contour interval = 5×10^{-5} m/sec^2)

at $(\lambda,\phi) = (79°,2°)$, $\left.\frac{\partial(\Delta U)}{\partial r}\right|_{max} = 1.03 \times 10^{-3}$ m/sec^2

at $(\lambda,\phi) = (121°,3°)$, $\left.\frac{\partial(\Delta U)}{\partial r}\right|_{min} = -0.66 \times 10^{-3}$ m/sec^2

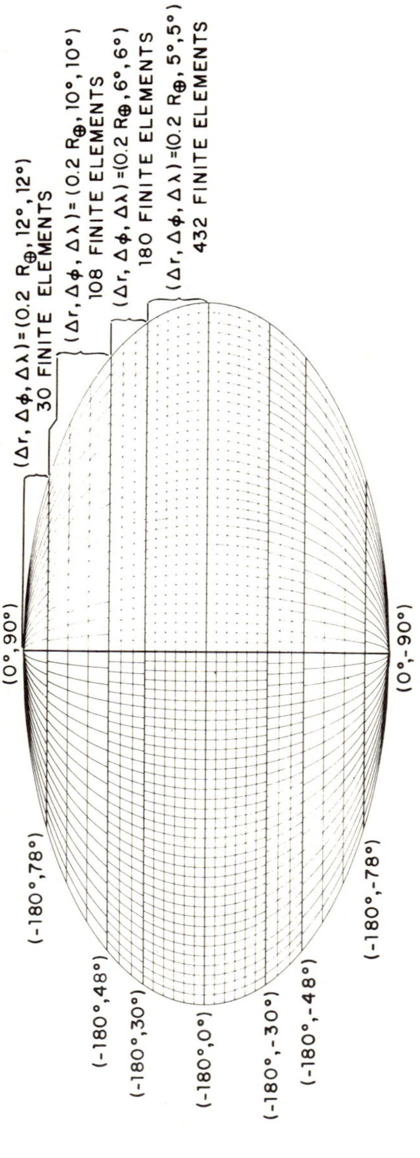

Figure 3.19 Distribution of Finite Elements for a 1500 Element 3rd Order Chebychev Model of the Geopotential

$$\text{GSH} = \begin{Bmatrix} G_S \\ G_E \\ G_R \end{Bmatrix}_{\substack{\text{Spherical} \\ \text{Harmonic} \\ \text{Series Model}}} \tag{3.71}$$

$$\text{GFE} = \begin{Bmatrix} G_S \\ G_E \\ G_R \end{Bmatrix}_{\substack{\text{Finite Element} \\ \text{Model}}} \tag{3.72}$$

was computed as were its RMS and maximum value for N = 23 in equations (3.44) and (3.46). We found that

$$\overline{\Delta G} = \frac{1}{N} \sum_{i=1}^{N} \Delta G_i \qquad = 0.000\ 000\ 03 \quad \text{m/sec}^2$$

$$\text{RMS}_{\Delta G} = \left\{\frac{1}{N} \sum_{i=1}^{N} \Delta G_i^2\right\}^{\frac{1}{2}} \qquad = 0.000\ 002 \quad \text{m/sec}^2$$

$$\Delta G_{MAX} \qquad = 0.000\ 008 \quad \text{m/sec}^2$$

$N = 360 \times 91 = 32760$ sample points ($1^\circ \times 1^\circ$ grid in the Northern hemisphere)

The above errors are the worst case errors arising generally in the most anomalous region near the earth's surface; analogous statistical analyses reveal the magnitude of the gravity modeling errors on the surface of a 1.03 R sphere are about one half of the above values while the errors are reduced by order of magnitude on the surface of the 1.2 R sphere.

This level of precision is probably satisfactory for integration of most missile trajectories and satellite orbits; typical single revolution position integration errors have been found to be on the order of 0 to 25 meters. Clearly, greater precision can be easily achieved, if desired, by decreasing the finite element size (thereby increasing the number of elements), or by increasing the order of local approximations (thereby decreasing the computational efficiency of the method).

The computational speed of the finite element model versus typical spherical harmonic recursion (ref. 3.15) favors the finite element model by about an order of magnitude. The computational picture is complicated, however, by virtue of the fact that random access retrieval of previously stored coefficient subsets is necessary. In this case, each component of acceleration requires a total of (20)(1500) = 30,000 coefficients to define the entire global family of gravity functions {although only twenty are used in each element}. Since the elements are large (hundreds of miles) compared to small errors (tens of miles) associated with two body or other simplified dynamic extrapolations, simple logic can be devised to bring into central memory several local sets of coefficients *before they are needed* and thereby hold the lost time during random access to a minimum. For the ballistic missile problem; most of the acceleration evaluations occur in two local regions (i.e., during atmospheric powered flight, and during re-entry); thus hundreds of acceleration evaluations are likely within a single finite element.

SUMMARY OF PROBLEM 3.7

The finite element approach to modeling the geopotential has been studied analytically and numerically. Many degrees of freedom exist in our approach to this problem. The specific finite element model developed and discussed here is not put forth as the optimal computational model of the geopotential. Rather, we believe the prototype finite model to be a representative finite element geopotential model which will probably be improved upon in future refinements of our approach. The computational speed advantages over spherical harmonic expansions is clear, however (primarily because a 23rd order expansion is locally replaced by a 3rd order expansion). The ultimate computational speed advantage depends upon the number of random accesses required to maintain the appropriate local coefficients in core memory. For a given trajectory/orbit integration, simulations done to date support the conclusion that an order of magnitude savings is achieved.

3.7 Remarks

The applications of this chapter were arranged in order of increasing complexity. It is reasonable to expect the interested reader to be able to reproduce the solutions of problems 3.2.1, 3.2.2 and 3.3 without engaging in overly elaborate computer programming. However, the more elaborate applications of §3.4, 3.5, and 3.6 probably entail too much of a computational burden unless the reader is specifically interested in pursuing applications closely related to these examples. However, it is anticipated that most readers, having gained computational and analytical experience from the examples of the first two chapters and elsewhere will profit greatly by a careful study of these applications. The results are presented in an essence-oriented narrative; the reader should be able to extract much that is to be learned from these examples without the necessity of reproducing the numerical results. However, for the reader interested in pursuing any of these applications in greater detail, adequate citations of pertinent literature are given.

The constraints imposed by the length of this text did not, however, permit an entirely self-contained and satisfactory development of the concepts (photogrammetric triangulation, gravitational potential theory, and weighting function approximation techniques) introduced in the applications of this chapter. It will likely prove useful for the interested reader to pursue these important subjects in the cited literature.

3.8 References for Chapter 3

3.1 Kelly, H.J., "Method of Gradients", chapter 6 of Optimization Techniques, Edited by Leitmann, Academic Press, pp 206-254.

3.2 Smithsonian Astrophysical Observatory Star Catalog, Smithsonian Institution, Washington, D.C., 1966.

3.3 Thompson, M.M., Manual of Photogrammetry, American Society of Photogrammetry, 3rd Ed., Vol. 1, p. 469.

3.4 Junkins, J.L., C.C. White III, and J.D. Turner, "A New System for Star Pattern Recognition/Spacecraft Attitude Determination", presented to the Flight Mechanics/Estimation Theory Symposium held at NASA Goddard Space Flight Center, Greenbelt, Md., October 27-28, 1976, proceedings in press.

3.5 Brown, D., 1968, "A Unified Lunar Control Network", Photogrammetric Engineering, 34:12, pp. 1272-1292.

3.6 Brown, D., 1970, "The Rigorous and Simultaneous Adjustment of Lunar Orbital Photography", Rome Air Development Center Report RADC-TR-70-274.

3.7 Blanton, J.N., and J.L. Junkins, "Dynamical Constraints in Satellite Photogrammetry", AIAA Paper No. 76-824 presented to the AIAA/AAS Astrodynamics Specialists Conference, San Diego, Ca., August 1976.

3.8 Blanton, J.N., "A New Formulation of the Free Motion of Triaxial Rigid Bodies and Applications to Satellite Photogrammetric Triangulation", Ph.D. Dissertation, Univ. of Va., Charlottesville, Va., August 1976.

3.9 Junkins, G. W. Miller, and J.R. Jancaitis, "A Weighting Function Approach to Modeling of Irregular Surfaces", Journal of Geophysical Research, V.78 No. 11, pp. 1794-1803, April 1973.

3.10 Jancaitis, J.R., and J.L. Junkins, "Mathematical Techniques for Automated Cartography", Final Report, Contract No. DAAK02-72-C-0256, U.S. Army Engineer Topographic Laboratories, Ft. Belvoir, Va., Feb. 1973.

3.11 Jancaitis, J.R. and J.L. Junkins, "Modeling and Contouring Irregular Surfaces Subject to Constraints", Final Report Contract No. DAAK02-73-C-0213, U.S. Army Engineer Topographic Laboratories, Ft. Belvoir, Va., July 1974.

3.12 Junkins, J.L. and J.R. Jancaitis, "Piecewise Continuous Surfaces: A Flexible Basis for Digital Mapping", Paper No. 74-201, Proceedings of the Fall 1974 Convention of the American Society of Photogrammetry, Washington D.C., Sept. 1974.

3.13 Heiskanen, W.A. and Moritz, H., Physical Geodesy, W.H. Freeman and co. 1967, ch. 1-3.

3.14 Pines S., "Uniform Representation of the Gravitational Potential and Its Derivatives", AIAA Journal, Vol. 11, Nov. 1973, pp. 1508-1521.

3.15 Sitzman, G.L., "A Numerical Technique to Calculate the Earth's Gravity and Its Gradient in Rectangular and Spherical Coordinates", Naval Weapons Laboratory Tech. Report, TR-2471, Feb. 1971, Dahlgren, Va.

3.16 Jancaitis, J.R. and J.L. Junkins, "Modeling N Dimensional Surfaces Using a Weighting Function Approach", Journal of Geophysical Research, Vol. 79, Aug. 1974, pp. 3361-3366.

3.17 Junkins, J.L. "Investigation of Finite-Element Representations of the Geopotential", AIAA Journal Vol 14, No. 6, pp. 803-808, June 1976.

3.18 Junkins, J.L. "Development of a Finite Element Model of the Earth's Gravity Field", Univ. of Va., Research Laboratories for the Engineering Sciences Rept. ESS-3538-105-75, Dec. 1975, Charlottesville, Va.

4

Survey of Ordinary Differential Equations

4.1 *Preliminary Remarks*

The mathematical models of most physical processes embodies one or more differential equations. A large fraction of practical problems is included if we restrict our attention to the case in which the state is the solution of a system of ordinary differential equations (odes). The differential equations usually arise quite naturally from application of fundamental principles (e.g., Newton's laws of motion) known to govern the particular dynamical system's behavior. In a significant fraction of the applications, it is possible to obtain explicit algebraic solutions of the system of differential equations, when this is possible, the results of the first three chapters may be immediately employed. If simple algebraic analytical solutions of the differential equations cannot be found, one need not (necessarily!) despair. In the present chapter, we overview analytical and numerical methods for solving systems of differential equations. Chapter 5 develops estimation methodology, and Chapter 6 documents several prototype applications, for situations in which the system's mathematical model embodies one or more ordinary differential equations.

4.2 *The State Space Approach*

We consider here dynamical systems whose behavior is governed by a system of differential equations of the form

$$\dot{X} = F(X, U, t) \tag{4.1}$$

where

$$X^t = \{x_1\ x_2\ \dots\ x_n\} = \text{the } \textit{state vector}$$

$$U^t = \{u_1\ u_2\ \dots\ u_m\} = \text{the } \textit{control vector}$$

$$F(X, U, t) = \begin{Bmatrix} f_1(x_1, x_2, \dots x_n; u_1, u_2, \dots, u_m; t) \\ \vdots \\ f_n(x_1, x_2, \dots x_n; u_1, u_2, \dots, u_m; t) \end{Bmatrix}$$

= a vector of n generally nonlinear functions of all n + m + 1 arguments.

and

$$(\dot{\ }) \equiv \frac{d}{dt}(\ ) .$$

In many cases, the system of differential equations do not naturally occur in the form of equation (4.1), but can be brought to this structure by a simple change of variables. To illustrate this point, we consider several simple examples.

Example 4.1: A Forced Linear Oscillator

The *natural* structure of the linear oscillator ODE is

$$m\ddot{x} + c\dot{x} + kx = f(t) \tag{4.2}$$

where m, k, and c are constants (*mass, stiffness,* and *damping*, respectively), and f(t) is an arbitrary but specified *forcing function*. The variable change

$$x_1 = x$$

$$x_2 = \dot{x}$$

leads immediately to the equivalent first order system

$$\begin{aligned} \dot{x}_1 &= x_2 \\ \dot{x}_2 &= -\frac{k}{m}x_1 - \frac{c}{m}x_2 + u(t) \end{aligned} \tag{4.3}$$

where $u(t) \equiv \frac{1}{m} f(t)$; equations (4.3) are now in the standard form (4.1).

Example 4.2: A Maneuverable Satellite

Application of Newton's second law and the law of gravitation leads to the system of three nonlinear differential equations

$$\ddot{x} = -\frac{GM}{r^3} x + u_1 \tag{4.4a}$$

$$\ddot{y} = -\frac{GM}{r^3} y + u_2 \tag{4.4b}$$

$$\ddot{z} = -\frac{GM}{r^3} z + u_3 \tag{4.4c}$$

where

$r^2 = x^2 + y^2 + z^2$,

GM = The earth's gravitational-mass constant.

u_i = The components of thrust-induced acceleration.

The variable change

$$x_1 = x, \quad x_2 = y, \quad x_3 = z$$

$$x_4 = \dot{x}, \quad x_5 = \dot{y}, \quad x_6 = \dot{z}$$

leads immediately to the equivalent six first order equations

$$\begin{aligned}
\dot{x}_1 &= x_4 \\
\dot{x}_2 &= x_5 \\
\dot{x}_3 &= x_6 \\
\dot{x}_4 &= -\ GM\ x_1 (x_1^2 + x_2^2 + x_3^2)^{-3/2} + u_1 \\
\dot{x}_5 &= -\ GM\ x_1 (x_1^2 + x_2^2 + x_3^2)^{-3/2} + u_2 \\
\dot{x}_6 &= -\ GM\ x_3 (x_1^2 + x_2^2 + x_3^2)^{-3/2} + u_3
\end{aligned} \tag{4.5}$$

which is in the standard form of equations (4.1).

Example 4.3: An nth Order Linear ode.

Suppose that the original equation is

$$\frac{d^n x}{dt^n} + a_n \frac{d^{n-1} x}{dt^{n-1}} + \ldots + a_2 \frac{dx}{dt} + a_1 x + b = u. \tag{4.6}$$

The following variable change

$$x_1 = x$$
$$x_2 = \frac{dx}{dt}$$
$$\vdots$$
$$x_n = \frac{d^{n-1}x}{dt^{n-1}}$$

leads immediately to the equivalent system of n first order equations

$$\begin{aligned} \dot{x}_1 &= x_2 \\ \dot{x}_2 &= x_3 \\ &\vdots \\ \dot{x}_{n-1} &= x_n \\ \dot{x}_n &= -b - a_1 x_1 - a_2 x_2 - \ldots - a_n x_n + u \end{aligned} \qquad (4.7)$$

which is in standard form of (4.1).

Perhaps the above emphasis upon writing systems of differential equations as first order equations seems misplaced, however, this will be seen not to be the case. Among developers of computer software for solution of ordinary differential equations, there is almost universal adoption of the standardized form of eqn. (4.1). Thus, in theoretical developments whose end products are likely to be implemented on a computer, adherence to this convention is justified on practical grounds. Perhaps a more important analytical justification of the state space notation is that its use permits the development of a unified *geometric theory* for analysis of the solution of nonlinear odes (ref. 4.4, ch.5).

4.3 Linear Dynamical Systems

The simplest class of systems described by equations (4.1) is the case in which the state and control variables appear linearly and the differential equations can be brought to the form

$$\dot{X}(t) = A(t)\ X(t) + B(t)\ U(t). \qquad (4.8)$$

A significant class of real world systems are directly and satisfactorily modeled by a system of odes of the form (4.8). In a large class of problems in which (4.8) does not satisfactorily describe the state itself, the *departure* of the actual state history from a reference known motion *may be* adequately described by (4.8). In the latter case, the actual system state dynamics may be governed by a highly nonlinear system of odes of the form (4.1); equation (4.8) may still adequately describe "departure motion" from a *nominal trajectory** whose integration is formally indicated as

$$X_N(t) = X_N(0) + \int_0^t F(\tau, X_N, U_N)\,d\tau. \tag{4.9}$$

The following linear differential equation

$$\frac{d}{dt}(\delta X(t)) \simeq A(t)\delta X(t) + B(t)\delta U(t), \tag{4.10}$$

Results from a Taylor's expansion of (4.1) about the nominal trajectory (4.9) at each instant in time, where

$$X(t) = \text{Actual state} = X_N(t) + \delta X(t), \tag{4.11}$$

$$U(t) = \text{Actual control} = U_N(t) + \delta U(t), \tag{4.12}$$

$$A(t) = \left.\frac{\partial f_i}{\partial x_j}\right|_N = n_x \; x \; n_x \text{ "State Jacobian" evaluated along (4.9)}, \tag{4.13}$$

$$B(t) = \left.\frac{\partial f_i}{\partial u_j}\right|_N = n_x \; x \; n_u \text{ "Control Jacobian" evaluated along (4.9)}. \tag{4.14}$$

Equation (4.10) can be integrated and then employed in equation (4.11) to approximate trajectories in a sufficiently small neighborhood of (4.9). Equation (4.10) amounts to a linearization of "Encke's Method" {see Herrick (1971)} which integrates the actual (nonlinear) "perturbative departure" of actual motion from reference motion.

*Here we consider the important special case that the known motion (from which departures are reckoned) is in fact a particular solution ("nominal trajectory") of the system (4.8). The more general circumstance is treated in ref. 4.4, §5.1 and 6.4.

Estimation theory based upon a linear differential equation of the form (4.8) is seen to be applicable (at least approximately) to a wide class of dynamical systems. In any given application to nonlinear problems, of course, one must realistically face the problems of choosing suitable nominal trajectories to linearize about, and analyzing the effects of errors introduced through the linearization.

After discussion of the standard approaches for solution of odes of the form (4.8), several specific examples will be considered to help the reader focus upon typical problems.

4.3.1 Homogeneous Linear Dynamical Systems

Consider the homogeneous matrix differential equation

$$\dot{X}(t) = A(t)X(t) \ , \ X(t_0) \text{ known.} \tag{4.15}$$

The standard approach for solving equations of the form (4.15) is to determine the "fundamental" or "state transition" matrix $\Phi(t,t_0)$ which "maps" the initial state into the current state as

$$X(t) = \Phi(t,t_0)X(t_0). \tag{4.16}$$

Before developing means for determining $\Phi(t,t_0)$, three important group properties of the transition matrix which follow from inspection of (4.16) are stated as

$$\Phi(t_0,t_0) = I \tag{4.17}$$

$$\Phi(t_0,t) = \Phi^{-1}(t,t_0) \tag{4.18}$$

$$\Phi(t_2,t_0) = \Phi(t_2, t_1)\Phi(t_1,t_0). \tag{4.19}$$

A differential equation for determining $\Phi(t,t_0)$ can be developed by substituting (4.16) into the right side of (4.15) and the derivative of (4.16) into the left side of (4.15) to obtain

$$\dot{\Phi}(t,t_0)X(t_0) = A(t)\Phi(t,t_0)X(t_0) \tag{4.20}$$

from which we conclude that the transition matrix satisfies the differential equation

$$\dot{\Phi}(t,t_0) = A(t)\Phi(t,t_0) \quad , \tag{4.21}$$

with the identity matrix (4.17) as the initial condition.

Only under ideal circumstances can a practical analytical solution of (4.21) be obtained; otherwise, numerical techniques can be employed to compute $\Phi(t,t_0)$. If (4.15) represents a linearization about a nominal trajectory {e.g., as does equation (4.10)}, then it is usually convenient to numerically integrate (4.21) simultaneously with the integration of the nominal trajectory (4.19). We now consider several standard approaches for extracting analytical exact or approximate solutions for $\Phi(t,t_0)$.

To develop one approach for solving (4.21), we rewrite it in integral form as

$$\Phi(t,t_0) = I + \int_{t_o}^{t} A(\tau_1)\Phi(\tau_1,t_0)d\tau_1 \quad , \tag{4.22}$$

which is a "matrix Volterra integral equation". We "casually note" that the integrand of (4.22) contains the left side; so it does not appear that any progress has been made writing (4.21) in integral form. One "might consider the wisdom" of substituting (4.22) *into its own integrand;* while this process may appear not only obscene, but futile, it does turn out to be profitable! For $\Phi(\tau_1,t_0)$ in the integrand of (4.22), we substitute from (4.22)

$$\Phi(\tau_1,t_0) = I + \int A(\tau_2)\Phi(\tau_2,t_0)d\tau_2 \tag{4.23}$$

to obtain

$$\Phi(t,t_0) = I + \int A(\tau_1)d\tau_1$$

$$+ \int A(\tau_1) \int A(\tau_2)\Phi(\tau_2,t_0)d\tau_2 d\tau_1. \tag{4.24}$$

One can now re-use (4.22) to write

$$\Phi(\tau_2,t_0) = I + \int_{t_o}^{t} A(\tau_3)\Phi(\tau_3,t_0)d\tau_3 \tag{4.25}$$

which, when substituted into the final integrand of (4.24) yields

$$\begin{aligned}\Phi(t,t_0) = I &+ \int_{t_o}^{t} A(\tau_1)d\tau_1 \\ &+ \int_{t_o}^{t} A(\tau_1) \int_{t_o}^{\tau_1} A(\tau_2)d\tau_2 d\tau_1 \\ &+ \int_{t_o}^{t} A(\tau_1) \int_{t_o}^{\tau_1} A(\tau_2) \int_{t_o}^{\tau_2} A(\tau_3)d\tau_3 d\tau_2 d\tau_1 \\ &+ \ldots\ldots\end{aligned} \tag{4.26}$$

This procedure is known as the Peano-Baker Method; as is shown by Ince (1926), uniform and absolute convergence is guaranteed. Whether or not this process is practical depends, of course, upon how difficult the elements of the A(t) are to integrate, and how quickly convergence occurs.

Considering an important special case A = a constant matrix, A can be brought from under all integrands of (4.26), we immediately find

$$\begin{aligned}\Phi(t,t_0) = I + A(t - t_0) &+ \frac{1}{2!} A \cdot A(t - t_0)^2 + \ldots. \\ &+ \frac{1}{n!}A^n(t - t_0)^n \quad + \cdots\end{aligned} \tag{4.27}$$

which is recognized to be the e^x series with the matrix $A(t - t_0)$ as the argument. For notational compactness, equation (4.27) is often written compactly as

$$\Phi(t,t_0) = e^{A(t - t_0)}, \ \underline{\text{for } A = \text{constant}}. \tag{4.28}$$

(This conclusion can be established in a number of ways (c.f., the methods of successive approximation on p. 215 of ref. 4.4); we will subsequently develop this result via a Taylor's series expansion.)

Thus, returning to equation (4.16), we see that the solution for constant A is

$$X(t) = e^{A(t - t_o)} X(t_0) \quad . \tag{4.29}$$

Consider the analogy of the matrix differential equation (4.15) with the scalar differential equation

$$\dot{x}(t) = a(t)x(t) \;, \; x(t_0) \text{ known.} \tag{4.30}$$

For the special case that a = constant, then the solution of (4.30) is

$$x(t) = x_0 e^{a(t - t_o)} \quad . \tag{4.31}$$

Thus, except for the constrained order of multiplication, the matrix solution (4.29) of (4.15) is completely analogous to the scalar solution (4.31) of (4.30) for constant coefficient matrices.

For the general case that $a \neq$ constant, the general solution of (4.30) is

$$x(t) = x_0 e^{\int_{t_o}^{t} a(\tau)\, d\tau} \tag{4.32}$$

One might naturally conjecture that the general solution of (4.15) is

$$X(t) = \Phi(t,t_0)X(t_0) = \left[e^{\int_{t_o}^{t} A(\tau)d\tau}\right]X(t_0) \;; \tag{4.33}$$

this conjecture turns out to be false, in general. To see under what conditions equation (4.33) *is* a correct solution of (4.15), note

$$\Phi = e^{\int A d\tau} = I + \{\int A d\tau\} + \frac{1}{2!}\{\int A d\tau\}\{\int A d\tau\} + \frac{1}{3!}\{\int A d\tau\}\{\int A d\tau\} + \ldots \tag{4.34}$$

$$\dot{\Phi} = 0 + A + \frac{1}{2!}\; \{A\}\{\int A d\tau\} + \{\int A d\tau\}\{A\}$$

$$+ \frac{1}{3!}\; \{A\}\{\int A d\tau\}\{\int A d\tau\} + \{\int A d\tau\}\{A\}\{\int A d\tau\}$$

$$+ \{\int A d\tau\}\{\int A d\tau\}\{A\} \quad + \ldots \tag{4.35}$$

and

$$A\Phi = A + \{A\}\{\int A d\tau\} + \frac{1}{2!}\{A\}\{\int A d\tau\}\{\int A d\tau\} + \dots \qquad (4.36)$$

Clearly, for (4.35) and (4.36) to be equal {which they must if (4.38) is a solution of (4.15)} then it is necessary that the "commutivity property"

$$\left[A(t)\right]\left[\int_{t_o}^{t} A(\tau)d\tau\right] = \left[\int_{t_o}^{t} A(\tau)d\tau\right]\left[A(t)\right] \qquad (4.37)$$

be satisfied; this property defines a very special class of matrices!

The conclusion is that the analogy between solutions of (4.16) and its scalar analog is not complete. The Peano-Baker solution (4.26) can be written in a shorthand notation as

$$\Phi(t,t_0) = I + \sum_{i=1}^{\infty} \ell_i \qquad (4.38)$$

where the integrals are defined as

$$\ell_1(t) = \int_{t_o}^{t} A(\tau_1)d\tau_1 \qquad (4.39)$$

$$\ell_2(t) = \int_{t_o}^{t} A(\tau_1) \int_{t_o}^{\tau_1} A(\tau_2)d\tau_2 d\tau_1 = \int_{t_o}^{t} A(\tau_1)\ell_1(\tau_1)d\tau_1 \qquad (4.40)$$

$$\ell_3(t) = \int_{t_o}^{t} A(\tau_1) \int_{t_o}^{\tau_1} A(\tau_2) \int_{t_o}^{\tau_2} A(\tau_3)d\tau_3 d\tau_2 d\tau_1 \qquad (4.41)$$

or

$$\ell_3(t) = \int_{t_o}^{t} A(\tau_1)\ell_2(\tau_1)d\tau_1$$

$$\vdots$$

$$\ell_n(t) = \int A(\tau_1)\ell_{n-1}^{(\tau_1)}d\tau_1 \qquad (4.42)$$

As an alternative to the Peano-Baker solution, consider the Taylor's Series

$$\Phi(t,t_0) = I + \sum_{n=1}^{\infty} \frac{(t - t_0)^n}{n!} \left. \frac{d^n\Phi}{dt^n} \right|_{t=t_0} \tag{4.43}$$

where the necessary partial derivatives are evaluated sequentially from the following equations:

IN GENERAL	EVALUATED INITIALLY
$\frac{d\Phi}{dt} = A\,\Phi$	$\left.\frac{d\Phi}{dt}\right\|_{t=t_0} = A(t_0)$
$\frac{d^2\Phi}{dt^2} = \frac{dA}{dt}\,\Phi + A\,\frac{d\Phi}{dt}$	$\left.\frac{d^2\Phi}{dt^2}\right\|_{t=t_0} = \left.\frac{dA}{dt}\right\|_{t=t_0} + A(t_0)\,A(t_0)$

$$\frac{d^n\Phi}{dt^n} = \sum_{i=1}^{n} \binom{n-1}{i-1} \left[\frac{d^{i-1}A}{dt^{n-i}}\right] \left[\frac{d^{i-1}\Phi}{dt^{i-1}}\right] \tag{4.44}$$

In particular, if A = constant, then equation (4.44) reduces to simply

$$\left. \frac{d^n\Phi}{dt^n} \right|_{t=t_0} = A^n \tag{4.45}$$

and equation (4.43) becomes

$$\Phi(t,t_0) = I + \sum_{n=1}^{\infty} \frac{(t-t_0)^n}{n!} A^n \equiv e^{A(t-t_0)} \tag{4.46}$$

In practice, if $A \neq$ constant, and the Peano-Baker or Taylor's Series prove too cumbersome (due to slow convergence or algebraic difficulties), then one must resort to a numerical solution of equation (4.21) or (4.15).

4.3.2 Linear State Variable Transformations

The matrix exponential for an arbitrary constant matrix is expensive to compute if one requires a large number of terms in equation (4.27). Often, one can carry out a coordinate transformation which "blasts this problem into trivia." Consider the introduction of a new state vector Y which is linearly related to X via

$$X = B\,Y \tag{4.47}$$

then

$$Y = B^{-1}\,X \tag{4.48}$$

and

$$\dot{Y} = B^{-1}\,\dot{X} \tag{4.49}$$

Now, substitution of the X-differential equation (4.15) yields

$$\dot{Y} = B^{-1}\,A\,X,$$

and substitution of (4.47) then yields the differential equation for Y as

$$\dot{Y} = \Lambda\,Y \tag{4.50}$$

where the new coefficient matrix is given by the similarity transformation

$$\Lambda = B^{-1}\,A\,B\ . \tag{4.51}$$

Now the unspecified B-matrix can often be judiciously chosen so that Λ is diagonal; more generally, Λ can be brought to a block diagonal form (the "Jordan Canonical Form"). If Λ is in fact diagonal, it is clear that the solution is trivial since (4.50) can be written as

$$\begin{Bmatrix} \dot{y}_1 \\ \dot{y}_2 \\ \vdots \\ \dot{y}_n \end{Bmatrix} = \begin{pmatrix} \lambda_1 & 0 & \dots & 0 \\ 0 & \lambda_2 & \dots & 0 \\ \vdots & \vdots & \ddots & \\ 0 & 0 & \dots & \lambda_n \end{pmatrix} \begin{Bmatrix} y_1 \\ y_2 \\ \vdots \\ y_n \end{Bmatrix} \tag{4.52}$$

or

$$\dot{y}_j = \lambda_j y_j, \qquad j = 1,2,\dots,n \tag{4.53}$$

and the solution is simply

$$y_j = y_j(t_0)e^{\lambda_j(t-t_0)}, \qquad j = 1,2,\dots,n \tag{4.54}$$

The solutions (4.54) can be written in state transition matrix form as

$$Y(t) = \Psi(t,t_0)\ Y(t_0) \tag{4.55}$$

where

$$\Psi(t,t_0) \equiv \begin{bmatrix} e^{\lambda_1(t-t_0)} & & \\ & e^{\lambda_2(t-t_0)} & \\ & & e^{\lambda_n(t-t_0)} \end{bmatrix} \tag{4.56}$$

Now, since

$$X = B\ Y$$

Then substitution of (4.55) and (4.48) yields

$$X(t) = B\ \Psi(t,t_0)\ B^{-1}\ X(t_0)\ ; \tag{4.57}$$

the state transition matrix for X is then clearly identified as

$$\Phi(t,t_0) \equiv B\ \Psi(t,t_0)\ B^{-1}\ . \tag{4.58}$$

Let us now see how to construct the elements of the $\{B\}$ and $\{\Lambda\}$ matrices. We require that the similarity transformation yield a diagonal $\{\Lambda\}$ matrix as

$$\Lambda = B^{-1} A B$$

or

$$B \Lambda = A B . \tag{4.59}$$

In detail, the equations (4.59) are

$$\begin{pmatrix} b_{11} & \cdots & b_{1n} \\ \vdots & & \\ b_{n1} & \cdots & b_{nn} \end{pmatrix} \begin{pmatrix} \lambda_1 & & 0 \\ & \ddots & \\ 0 & & \lambda_n \end{pmatrix} = \begin{pmatrix} a_{11} & \cdots & a_{1n} \\ \vdots & \ddots & \vdots \\ a_{n1} & \cdots & a_{nn} \end{pmatrix} \begin{pmatrix} b_{11} & \cdots & b_{1n} \\ \vdots & \ddots & \vdots \\ b_{n1} & \cdots & b_{nn} \end{pmatrix} . \tag{4.60}$$

Equating the jth column resulting from the matrix product on the LHS of (4.60) to the jth column on the RHS yields

$$\lambda_j \begin{Bmatrix} b_{1j} \\ b_{2j} \\ \vdots \\ b_{nj} \end{Bmatrix} = \{A\} \begin{Bmatrix} b_{1j} \\ b_{2j} \\ \vdots \\ b_{nj} \end{Bmatrix} \qquad j = 1,2,\ldots,n. \tag{4.61}$$

Thus, the conclusion is that the diagonal elements of $\{\Lambda\}$ *are the eigenvalues of* $\{A\}$, *and the columns of the required matrix* $\{B\}$ *are the corresponding eigenvectors of* $\{A\}$. The λ's are the n roots of the characteristic equation

$$\det (\lambda I - A) = 0 \rightarrow \lambda_1, \lambda_2, \ldots, \lambda_n . \tag{4.62}$$

Upon determining λ_i's from (4.62), the b_{ij}'s are determined (to within an arbitrary multiplicative constant for each column) from (4.61). For the most common case that the n λ's satisfying (4.62) are distinct, then independent columns of B can always be found to satisfy (4.62). For the case that (4.62) has multiple roots, it is not always possible to find independent columns of B from (4.61) which will guarantee (4.51) to be diagonal. The difficulties encountered for repeated eigenvalues are not always trivial to resolve; see references 4.3 and 4.4 for a more detailed treatment of this subject.

4.3.3 Forced Linear Systems

We now direct our attention to the inhomogeneous differential equation

$$\dot{X}(t) = A(t)X(t) + B(t)U(t). \tag{4.63}$$

Using Lagrange's method of *variation of parameters*, we assume a solution of (4.68) having the form

$$X(t) = \Phi(t,t_0)G(t), \; G(t_0) = X(t_0), \tag{4.64}$$

where G(t) is an n_x *x* 1 matrix of unknown functions to be determined and $\Phi(t,t_0)$ is the homogeneous transition matrix. Differentiating (4.64), we obtain

$$\dot{X}(t) = \dot{\Phi}(t,t_0)G(t) + \Phi(t,t_0)\dot{G}(t) \tag{4.65}$$

which, upon substitution of equation (4.21) for $\dot{\Phi}(t,t_0)$, becomes

$$\dot{X}(t) = \Phi(t,t_0)\dot{G}(t) + A(t)\Phi(t,t_0)G(t) \; . \tag{4.66}$$

Substituting (4.64) and (4.66) into (4.63) yields

$$\Phi(t,t_0)\dot{G}(t) + A(t)\Phi(t,t_0)G(t) = A(t)\Phi(t,t_0)G(t) + B(t)U(t), \tag{4.67}$$

from which

$$\dot{G}(t) = \Phi^{-1}(t,t_0)B(t)U(t) \; , \tag{4.68}$$

which we integrate to obtain {noting $G(t_0) \equiv X(t_0)$, for (4.64) to be valid at $t = t_0$}

$$G(t) = X(t_0) + \int_{t_o}^{t} \Phi^{-1}(\tau,t_0)B(\tau)U(\tau)d\tau \; . \tag{4.69}$$

Therefore, the general solution of (4.63) is

$$X(t) = \Phi(t,t_0)X(t_0) + \Phi(t,t_0)\int_{t_0}^{t} \Phi^{-1}(\tau,t_0)B(\tau)U(\tau)d\tau \qquad (4.70)$$

Application of (4.18) allows the integrand to be written as

$$\Phi^{-1}(\tau,t_0) = \Phi(t_0,\tau) \qquad (4.71)$$

and, since (4.19) then

$$\Phi^{-1}(\tau,t_0) = \Phi(t_0,\tau)$$

or

$$\Phi^{-1}(\tau,t_0) = \Phi(t_0,t)\Phi(t,\tau) \qquad (4.72)$$

or

$$\Phi^{-1}(\tau,t_0) = \Phi^{-1}(t,t_0)\Phi(t,\tau)$$

which, when substituted into (4.70) yields

$$X(t) = \Phi(t,t_0)X(t_0) + \int_{t_0}^{t} \Phi(t,\tau)B(\tau)U(\tau)d\tau \qquad (4.73)$$

as the final form of the solution of (4.63) for arbitrary B(t), A(t), and U(t), equation, (4.73) must typically be solved numerically. We now consider several illustrative examples.

Example 4.4: Projectile Motion

Consider the motion of a projectile in a constant gravity field, the equations of motion are

$$\begin{aligned} \ddot{x} &= 0 \\ \ddot{y} &= 0 \\ \ddot{z} &= -g \end{aligned} \qquad (4.74)$$

which integrate immediately to

$$\begin{aligned} \dot{x} &= \dot{x}_0 \\ \dot{y} &= \dot{y}_0 \\ \dot{z} &= \dot{z}_0 - g(t - t_0) \end{aligned} \qquad (4.75)$$

and

$$
\begin{aligned}
x &= x_0 + (t - t_0)\dot{x}_0 \\
y &= y_0 + (t - t_0)\dot{y}_0 \\
z &= z_0 + (t - t_0)\dot{z}_0 - \tfrac{1}{2}(t - t_0)^2 g
\end{aligned}
\tag{4.76}
$$

Alternatively, we could have employed the variable change $\{x_1 = x,\ x_2 = y,\ x_3 = z,\ x_4 = \dot{x},\ x_5 = \dot{y},\ x_6 = \dot{z},\ u = -g\}$ so that (4.74) is brought to the standard state space form

$$
\underset{(6 \times 1)}{\dot{X}} = \underset{(6 \times 1)}{F(t,X,u)}
\tag{4.77}
$$

where

$$
\begin{aligned}
f_1(t,x) &= x_4 \\
f_2(t,x) &= x_5 \\
f_3(t,x) &= x_6 \\
f_4(t,x) &= 0 \\
f_5(t,x) &= 0 \\
f_6(t,x) &= -g
\end{aligned}
\tag{4.78}
$$

which integrates to the same results (4.76) and (4.77). In this case (4.77) is linear and can also be written as

$$
\dot{X} = AX + BU
\tag{4.79}
$$

or

$$
\begin{bmatrix} \dot{x}_1 \\ \dot{x}_2 \\ \dot{x}_3 \\ \dot{x}_4 \\ \dot{x}_5 \\ \dot{x}_6 \end{bmatrix}
=
\begin{bmatrix}
0 & 0 & 0 & 1 & 0 & 0 \\
0 & 0 & 0 & 0 & 1 & 0 \\
0 & 0 & 0 & 0 & 0 & 1 \\
0 & 0 & 0 & 0 & 0 & 0 \\
0 & 0 & 0 & 0 & 0 & 0 \\
0 & 0 & 0 & 0 & 0 & 0
\end{bmatrix}
\begin{bmatrix} x_1 \\ x_2 \\ x_3 \\ x_4 \\ x_5 \\ x_6 \end{bmatrix}
+
\begin{bmatrix}
0 & 0 & 0 & 0 & 0 & 0 \\
0 & 0 & 0 & 0 & 0 & 0 \\
0 & 0 & 0 & 0 & 0 & 0 \\
0 & 0 & 0 & 0 & 0 & 0 \\
0 & 0 & 0 & 0 & 0 & 0 \\
0 & 0 & 0 & 0 & 0 & 1
\end{bmatrix}
\begin{bmatrix} 0 \\ 0 \\ 0 \\ 0 \\ 0 \\ -g \end{bmatrix} .
\tag{4.80}
$$

Notice, by inspection of (4.75) and (4.76), that the state transition matrix is

$$
\Phi(t,t_0) =
\begin{bmatrix}
1 & 0 & 0 & (t - t_0) & 0 & 0 \\
0 & 1 & 0 & 0 & (t - t_0) & 0 \\
0 & 0 & 1 & 0 & 0 & (t - t_0) \\
0 & 0 & 0 & 1 & 0 & 0 \\
0 & 0 & 0 & 0 & 1 & 0 \\
0 & 0 & 0 & 0 & 0 & 1
\end{bmatrix} .
\tag{4.81}
$$

The homogeneous solution is then

$$X(t) = \Phi(t,t_0)X(t_0) \cdot \tag{4.82}$$

The "forced" solution (including gravity) follows from (4.73) as

$$X(t) = \Phi(t,t_0)X(t_0) + \int_{t_o}^{t} \Phi(t,\tau)B(\tau)U(\tau)d\tau \tag{4.83}$$

or, substituting (4.78) into (4.83)

$$\begin{Bmatrix} x_1(t) \\ x_2(t) \\ x_3(t) \\ x_4(t) \\ x_5(t) \\ x_6(t) \end{Bmatrix} = \begin{Bmatrix} 1 & 0 & 0 & (t - t_0) & 0 & 0 \\ 0 & 1 & 0 & 0 & (t - t_0) & 0 \\ 0 & 0 & 1 & 0 & 0 & (t - t_0) \\ 0 & 0 & 0 & 1 & 0 & 0 \\ 0 & 0 & 0 & 0 & 1 & 0 \\ 0 & 0 & 0 & 0 & 0 & 1 \end{Bmatrix} \begin{Bmatrix} x_1(t_0) \\ x_2(t_0) \\ x_3(t_0) \\ x_4(t_0) \\ x_5(t_0) \\ x_6(t_0) \end{Bmatrix} +$$

$$\int_{t_o}^{t} \begin{Bmatrix} 1 & 0 & 0 & (t-\tau) & 0 & 0 & 0 & 0 & 0 & 0 & 0 & 0 \\ 0 & 1 & 0 & 0 & (t-\tau) & 0 & 0 & 0 & 0 & 0 & 0 & 0 \\ 0 & 0 & 1 & 0 & 0 & (t-\tau) & 0 & 0 & 0 & 0 & 0 & 0 \\ 0 & 0 & 0 & 1 & 0 & 0 & 0 & 0 & 0 & 0 & 0 & 0 \\ 0 & 0 & 0 & 0 & 1 & 0 & 0 & 0 & 0 & 0 & 0 & 0 \\ 0 & 0 & 0 & 0 & 0 & 1 & 0 & 0 & 0 & 0 & 0 & 1 \end{Bmatrix} \begin{Bmatrix} 0 \\ 0 \\ 0 \\ 0 \\ 0 \\ -g \end{Bmatrix} d\tau$$

or

$$\begin{Bmatrix} x_1(t) \\ x_2(t) \\ x_3(t) \\ x_4(t) \\ x_5(t) \\ x_6(t) \end{Bmatrix} = \begin{Bmatrix} x_1(t_0) + (t-t_0)x_4(t_0) \\ x_2(t_0) + (t-t_0)x_5(t_0) \\ x_3(t_0) + (t-t_0)x_6(t_0) \\ x_4(t_0) \\ x_5(t_0) \\ x_6(t_0) \end{Bmatrix} + \int_{t_o}^{t} \begin{Bmatrix} 0 \\ 0 \\ -g(t-\tau) \\ 0 \\ 0 \\ -g \end{Bmatrix} d\tau$$

or

$$\begin{Bmatrix} x_1(t) \\ x_2(t) \\ x_3(t) \\ x_4(t) \\ x_5(t) \\ x_6(t) \end{Bmatrix} = \begin{Bmatrix} x_1(t_0) + (t-t_0)x_4(t_0) \\ x_2(t_0) + (t-t_0)x_5(t_0) \\ x_3(t_0) + (t-t_0)x_6(t_0) - \tfrac{1}{2}g(t-t_0)^2 \\ x_4(t_0) \\ x_5(t_0) \\ x_6(t_0) - g(t-t_0) \end{Bmatrix} \tag{4.84}$$

which verify (again) the results (4.75) and (4.76) and demonstrates the equivalence between the preceding results of this section and conventional integration of the differential equations.

Example 4.5: Satellite Orbit Integration

Consider a satellite moving through a general gravity field having a known potential function $V(x,y,z)$, the equations of translational motion of the satellite are

$$\begin{aligned}\ddot{x} &= \frac{\partial V}{\partial x}(x,y,z) + X(x,y,z,\dot{x},\dot{y},\dot{z},\ldots)\\ \ddot{y} &= \frac{\partial V}{\partial y}(x,y,z) + Y(x,y,z,\dot{x},\dot{y},\dot{z},\ldots)\\ \ddot{z} &= \frac{\partial V}{\partial z}(x,y,z) + Z(x,y,z,\dot{x},\dot{y},\dot{z},\ldots)\end{aligned} \tag{4.85}$$

where $\{X,Y,Z\}$ are the components of all accelerations *not* due to the gravity model (e.g., thrust, drag, solar radiation pressure,...).

Equations (4.85) can be transformed to a system of six first order odes via the variable changes $\{x_1 = x,\ x_2 = y,\ x_3 = z,\ x_4 = \dot{x},\ x_5 = \dot{y},\ x_6 = \dot{z}\}$:

$$\dot{X} = F(t,X) \tag{4.86}$$

where

$$\begin{aligned}f_1(t,X) &= x_4\\ f_2(t,X) &= x_5\\ f_3(t,X) &= x_6\\ f_4(t,X) &= \frac{\partial V}{\partial x_1} + X\\ f_5(t,X) &= \frac{\partial V}{\partial x_2} + Y\\ f_6(t,X) &= \frac{\partial V}{\partial x_3} + Z\ .\end{aligned} \tag{4.87}$$

Suppose a nominal trajectory of (4.86) is integrated as

$$X_N(t) = X_N(t_o) + \int_{t_o}^{t} F(\tau,X_n)d\tau\ . \tag{4.88}$$

The "variational equations" resulting from linearizing (4.86) about (4.88) are

$$\frac{d}{dt}(\delta X(t)) \equiv A(t)\delta X(t) \tag{4.89}$$

where

$$A(t) = \left(\left.\frac{\partial f_i}{\partial x_j}\right|_N \right) \tag{4.90}$$

and where the elements of A(t) are

$$\left.\frac{\partial f_1}{\partial x_i}\right|_N = 0 \qquad i = 1,2,3,5,6$$

$$\left.\frac{\partial f_1}{\partial x_4}\right|_N = 1$$

$$\left.\frac{\partial f_2}{\partial x_i}\right|_N = 0 \qquad i = 1,2,3,4,6$$

$$\left.\frac{\partial f_2}{\partial x_5}\right|_N = 1$$

$$\left.\frac{\partial f_3}{\partial x_i}\right|_N = 0 \qquad i = 1,2,3,4,5$$

$$\left.\frac{\partial f_3}{\partial x_6}\right|_N = 1 \tag{4.91}$$

$$\left.\frac{\partial f_4}{\partial x_i}\right|_N = \frac{\partial}{\partial x_i}\left.\frac{\partial V}{\partial x_1}\right|_N + \left.\frac{\partial X}{\partial x_i}\right|_N \qquad i = 1,2,3$$

$$\left.\frac{\partial f_4}{x_j}\right|_N = \left.\frac{\partial X}{\partial x_j}\right|_N \qquad j = 4,5,6$$

$$\left.\frac{\partial f_5}{\partial x_i}\right|_N = \frac{\partial}{\partial x_i}\left.\frac{\partial V}{\partial x_2}\right|_N + \left.\frac{\partial Y}{\partial x_i}\right|_N \qquad i = 1,2,3$$

$$\left.\frac{\partial f_5}{\partial x_j}\right|_N = \left.\frac{\partial Y}{\partial x_j}\right|_N \qquad j = 4,5,6$$

$$\left.\frac{\partial f_6}{\partial x_i}\right|_N = \frac{\partial}{\partial x_i}\left.\frac{\partial V}{\partial x_3}\right|_N + \left.\frac{\partial Z}{\partial x_i}\right|_N \qquad i = 1,2,3$$

$$\left.\frac{\partial f_6}{\partial x_j}\right|_N = \left.\frac{\partial Z}{\partial x_j}\right|_N \qquad j = 4,5,6.$$

If the transition matrix is required (e.g., to form partials for differential correction), then

$$\dot{\Phi}(t,t_0) = A(t)\Phi(t,t_0), \quad \Phi(t_0,t_0) = I \tag{4.92}$$

would typically be integrated simultaneously with the nominal trajectory (4.88). The actual state is then approximated as

$$X(t) \cong X_N(t) + \Phi(t,t_0)\delta X(t_0) \tag{4.93}$$

to within the validity of (4.89).

4.4 *Nonlinear Dynamical Systems*

4.4.1 *Overview*

We now consider the circumstance in which the original system of differential equations is nonlinear and can be brought to the standard form

$$\dot{X} = F(t, X, \ldots). \tag{4.94}$$

A significant minority of the nonlinear systems of differential equations encountered in applications can be solved for an exact analytical solution. Consider the following example.

Example 4.6: The Eliptic Two-Body Problem

The two-body relative acceleration equations are the coupled nonlinear system

$$\ddot{x} = -\frac{\mu}{r^3}x \qquad \ddot{z} = -\frac{\mu}{r^3}z$$
$$\ddot{y} = -\frac{\mu}{r^3}y \tag{4.95}$$

where

$\mu = G(M_1 + M_2)$ = gravitational-mass constant

$r^2 = x^2 + y^2 + z^2$

Herrick (1973) establishes that the analytical solution of (4.95) is the following equations in-order-of solution:

$$r_o^2 = x_o^2 + y_o^2 + z_o^2$$

$$V_o^2 = \dot{x}_o^2 + \dot{y}_o^2 + \dot{z}_o^2$$

$$D_o = x_o\dot{x}_o + y_o\dot{y}_o + z_o\dot{z}_o$$

$$\frac{1}{a} = \frac{2}{r_o} - V_o^2/\mu$$

$$c_o = 1 - r_o/a$$

solve for ϕ (using Newton's method) from

$$\mu^{\frac{1}{2}}(t - t_o) = a^{3/2}\left(\phi - (1 - \frac{r_o}{a})\sin\phi + \frac{D_o}{(\mu a)^{\frac{1}{2}}}(1 - \cos\phi)\right)$$

$$f = 1 - a(1 - \cos\phi)/r_o$$

$$g = (t - t_o) - a^{3/2}(\phi - \sin\phi)/\mu^{\frac{1}{2}}$$

$$r = a(1 - c_o\cos\phi) - D_o\left(\frac{a}{\mu}\right)^{\frac{1}{2}}\sin\phi$$

$$\dot{f} = -(rr_o)^{-1}(\mu a)^{\frac{1}{2}}\sin\phi$$

$$\dot{g} = 1 - a(1 - \cos\phi)/r$$

then the solution of (4.95) is

$$\begin{Bmatrix} x(t) \\ y(t) \\ z(t) \end{Bmatrix} = f \begin{Bmatrix} x_o \\ y_o \\ z_o \end{Bmatrix} + g \begin{Bmatrix} \dot{x}_o \\ \dot{y}_o \\ \dot{z}_o \end{Bmatrix} \tag{4.96a}$$

$$\begin{Bmatrix} \dot{x}(t) \\ \dot{y}(t) \\ \dot{z}(t) \end{Bmatrix} = \dot{f} \begin{Bmatrix} x_o \\ y_o \\ z_o \end{Bmatrix} + \dot{g} \begin{Bmatrix} \dot{x}_o \\ \dot{y}_o \\ \dot{z}_o \end{Bmatrix} . \tag{4.96b}$$

Unfortunately, only a minority of the nonlinear differential systems have known analytical solutions and no standardized methods exist for finding *exact analytical solutions*.

For the *perturbation class* of nonlinear systems whose differential equations can be brought to the form

$$\dot{X} = F(t,X) + G(t,X,\ldots) \tag{4.97}$$

in which the "unperturbed" *generating system*

$$\dot{Z} = F(t,Z)$$

has a known analytical solution, and

$$\left\| G(t,X,\ldots) \right\| \ll \left\| F(t,X) \right\| \tag{4.98}$$

for all t and X of practical interest, numerous methods are available for construction of *approximate* analytical solutions. In particular, a large family of perturbation theories have been developed for small perturbations of equations (4.95). The interested reader is referred to references 4.4, 4.5, and 4.6 for development of basic perturbation methods which are not developed herein due to space limitations. Let us remark, how-ever, that the perturbation approach suffers from one fundamental drawback; for each specification of the functions F and G in equations (4.97), lengthly algebraic developments must be carried through to obtain an *approximate* solution. In many cases the practical constraints imposed by "having but one life to give" and the desirability of constructing general-purpose algorithms make the analytical perturbation approach unattractive.

On the other hand, general-purpose numerical methods exist which are routinely employed to solve a wide variety of highly nonlinear systems of the form (4.94) with excellent, near-arbitrary control over precision of the solution. Due to the broad applicability of the numerical approach to solving non-linear odes, we summarize two of the simpler single-step class of algorithms in the following subsections.

4.4.2 Analytical Continuation Methods

Consider a single scalar differential equation of the form

$$\dot{x} = f(t,x) \quad , \quad x_o = x(t_o) \text{ known;} \tag{4.99}$$

it is desired to determine the "trajectory" $x(t)$ for $t_o \leq t \leq t_f$. For the common event that the right-hand-side of (4.99) is a continuous function with continuous derivatives, one straight forward approach to approximating $x(t)$ is to expand x as a Taylor's series about $t = t_o$ as

$$x(t) = x_o + \sum_{n=1}^{N} \left.\frac{d^n x}{dt^n}\right|_o \frac{(t-t_o)^n}{n!} + O(t-t_o)^{N+1} \tag{4.100}$$

where the necessary derivatives in (4.100) are determined from (4.99) as the sequence of equations

$$\begin{aligned} \left.\frac{dx}{dt}\right|_o &= f(t_o,x_o) \\ \left.\frac{d^2x}{dt^2}\right|_o &= \left.\frac{\partial f}{\partial t}\right|_o + \left.\frac{\partial f}{\partial x}\right|_o \left.\frac{dx}{dt}\right|_o \\ &\vdots \end{aligned} \tag{4.101}$$

Observe that the derivatives (4.101) have the structure

$$\frac{d^n x}{dt^n} = \text{function}\left(x, \frac{dx}{dt}, \frac{d^2x}{dt^2}, \ldots, \frac{d^{n-1}x}{dt^{n-1}} \right) \tag{4.102}$$

thus the nth derivative can, in principle, be calculated by evaluating equations (4.101) *sequentially*. However, for typical $f(t,x)$, the lengthy algebra implicit in (4.101) is often sufficiently intimidating to insure $N \leq 5$. Thus, the truncated series (4.100) are not attractive for large $(t-t_o)$.

Analytical continuation is a process of successive application of (4.100) in the form

$$x(t_{k+1}) = x(t_k) + \sum_{n=1}^{N} \left.\frac{d^n x}{dt^n}\right|_{t_k} \frac{h^n}{n!} + O(h^{N+1}) \tag{4.103}$$

where

$$h = t_{k+1} - t_k = \textit{step size}$$

N and h are chosen so that the terms of order higher than N contribute insignificantly to the approximation (4.103). The basic idea can probably be efficiently illustrated by a specific example (after Milne).

Example 4.7

Consider the scalar differential equation

$$\dot{x} = 1 - x^2 \quad , \quad x(0) = 0 \tag{4.104}$$

Successive differentiations of (4.104) yield the first five derivatives

$$\begin{aligned}
\frac{dx}{dt} &= 1 - x^2 \\
\frac{d^2x}{dt^2} &= -2x\,\frac{dx}{dt} \\
\frac{d^3x}{dt} &= -2\left(\frac{dx}{dt}\right)^2 - 2x\,\frac{d^2x}{dt^2} \\
\frac{d^4x}{dt^4} &= -6\left(\frac{dx}{dt}\right)\frac{d^2x}{dt^2} - 2x\left(\frac{d^3x}{dt^3}\right) \\
\frac{d^5x}{dt^5} &= -6\left(\frac{d^2x}{dt^2}\right)^2 - 8\left(\frac{dx}{dt}\right)\frac{d^3x}{dt^3} - 2x\left(\frac{d^4x}{dt^4}\right) \\
&\ \vdots
\end{aligned} \tag{4.105}$$

Choosing the initial conditions x = 0 at t = 0 with N = 5 and the step size h = 0.3 = constant results in the numerical solution tabulated in table 4.1. The column E is the absolute error difference between equation (4.103) and the exact analytical solution x = tanh t. In the example under discussion, the step size and order are specified, in a real application the selection of step size (and/or order) should be determined as discussed in §4.4.4.

Table 4.7 Analytical Continuation Solution of $\dot{x} = 1 - x^2$

t_k	x_k	$\left.\frac{dx}{dt}\right\|_k$	$\left.\frac{d^2x}{dt^2}\right\|_k$	$\left.\frac{d^3x}{dt^3}\right\|_k$	$\left.\frac{d^4x}{dt^4}\right\|_k$	$\left.\frac{d^5x}{dt^5}\right\|_k$	E
0.0	0.00000	1.00000	0.0000	-2.000	0.00	16.0	
0.3	0.29132	0.91513	-0.5332	-1.364	3.72	6.1	0.00001
0.6	0.53710	0.71152	-0.7643	-0.192	3.47	-6.1	0.00005
0.9	0.71635	0.48684	-0.6975	0.525	1.29	6.8	0.00005
1.2	0.83368	0.30498					0.00003

The above developments apply, in principle, to systems of non-linear differential equations. The structure of the algorithm is identical, if equation (4.99) is considered a vector-matrix equation; of course the algebra must then be considered to be matrix algebra.

Analytical continuation is not routinely applied, due to the necessity of deriving and calculating with the often cumbersome derivative equations (if the derivatives exist and are feasible to derive). A large variety of numerical methods exist which do not require analytical derivation of derivative expressions, and are applicable to equations of the form (4.94) for which derivatives above first order need not be continuous. The most widely applied such approach is summarized below in §4.4.3.

4.4.3 Runge - Kutta Methods

Given a system of differential equations of the form

$$\dot{X} = F(t,X,\ldots)\ , \quad X(t_o) \text{ given} \tag{4.94}$$

the analytical continuation algorithm for this system has the form

$$X(t_{k+1}) = X(t_k) + \sum_{n=1}^{N} \left.\frac{d^nX}{dt^n}\right|_{t_k} h^n/n! \tag{4.106}$$

The fourth order Runge-Kutta algorithm has the structure

$$X(t_{k+1}) = X(t_k) + \frac{1}{6}(\Delta_1 X + 2\Delta_2 X + 2\Delta_3 X + \Delta_4 X) \tag{4.107}$$

where the X-change approximations averaged in (4.107) are

$$\begin{aligned}
\Delta_1 X &= hF\{t_k, X(t_k), \ldots\} = \text{"left" approximation} \\
\Delta_2 X &= hF\{t_k + h/2, X(t_k) + \Delta_1 X/2, \ldots\} = \text{"first mid-point approximation"} \\
\Delta_3 X &= hF\{t_k + h/2, X(t_k) + \Delta_2 X/2, \ldots\} = \text{"second mid-point" approximation} \\
\Delta_4 X &= hF\{t_{k+1}, X(t_k) + \Delta_3 X \ldots\} = \text{"right" approximation}
\end{aligned} \tag{4.108}$$

This widely-used algorithm is shown by Milne to be of fourth order in h, and requires only four evaluations (4.108) of the right-hand side of the given differential equations (4.94). Thus no special purpose analytical developments need be carried out for each application. It is necessary, however, to carefully determine the step size (see §4.4.4). In many applications, it is useful to employ variable step sizes with associated logic to monitor local truncation errors.

Exercise 4.8

Apply the Runge-Kutta algorithm of eqns. (4.107) and (4.108) to the differential equation $\dot{x} = 1 - x^2$, $x_o = 0$. Vary the step size as h = 0.1, 0.2, 0.3, 0.4, 0.5 and compare step-by-step solution with the exact solution x = tanh t, and with the analytical continuation solution (example 4.7); for $0 \leq t \leq 1$.

4.4.4 *Closure Error Analysis*

The accumulation of errors in numerical solution of differential equations requires careful evaluation in practical applications. Rigorous error analysis are very difficult, unless the differential equations being solved have an analytical solution (in which use of numerical methods is redundant). However, a number of approximate error analysis procedures have been developed which allow routine, confident, and accurate numerical solution of differential equations.

The errors implicit in the solution $\{X(t_{k+1})\}$ at any integration step depends primarily upon the following factors

* The nonlinearity of $F(t,X,\ldots)$
* The smallness of h
* The order of the integration process
* The number of significant figures carried in the computations

A simple *closure test* useful to select an efficient step-size is based upon testing the magnitude of the relative closure error

$$\Delta_{12} = \sqrt{\sum_{i=1}^{n} \left(\frac{x_{i2}(t_f) - x_{i1}(t_f)}{K_i} \right)^2} \tag{4.109}$$

where

$$X_1(t_f) = \begin{Bmatrix} x_{11}(t_f) \\ x_{21}(t_f) \\ \vdots \\ x_{n1}(t_f) \end{Bmatrix}$$ = the state approximation resulting from step size h_1 {using, as examples, analytical continuation (4.106) or Runge-Kutta (4.107)}.

$$X_2(t_f) = \begin{Bmatrix} x_{12}(t_f) \\ x_{22}(t_f) \\ \vdots \\ x_{n2}(t_f) \end{Bmatrix}$$ = the state approximation resulting from step size $h_2 < h_1$ (observe, unless $t_f - t_o$ is an integer multiple of h_i, that a smaller final step will be required).

K_i = nondimensionalizing constants, usually taken as "nominal" or "typical" values of $x_i(t_f)$.

If $\chi(t_f)$ were replaced by the (usually unknown) true state, then eqn. (4.109) yields an *absolute* closure error. Application of analytical continuation, Runge-Kutta, and other methods to problems with known analytical solution reveal that closure error versus step size has a functional dependence typified by the graph of figure 4.1.

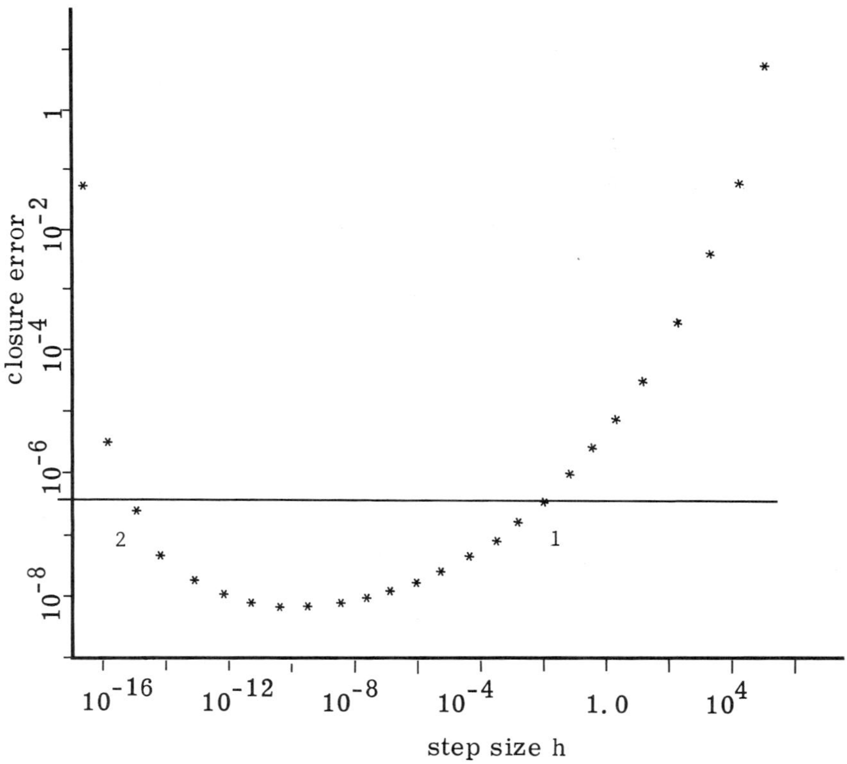

Figure 4.1 Closure Error vs Integration Step Size

Intuitively, one might expect that as h goes to zero, the approximations implicit in the particular solution method should go to zero also. This is generally true. Unfortunately, the number of arithmetic operations increase linearly with reductions in step size (since more steps are required to span the interval $t_f - t_o$). Since each arithmetic operation introduces errors in the final significant figure carried in the particular computer being used, the accumulation of these errors becomes most significant as the number of required steps becomes large. It is also important that small integration steps lead to long computer run time. Clearly, if in figure 4.1, the desired tolerance is 10^{-6}, then point 1 ($h = 10^{-1}$) is the optimum step size.

As an alternative approach to the above closure error approach, a common procedure is to compare a higher order integration process with a lower order process for a fixed step size. The relative error between the two integrations can then be employed as a basis for accepting or modifying the step size.

Exercise 4.9

From the calculations of Exercise 4.8, construct a closure error graph analogous to Figure 4.1.

4.5 Parametric Differentiation

Estimation or optimization algorithms are often applied to systems whose state is governed by a system of equations of the form

$$\dot{X} = F(t,X,P) \tag{4.110}$$

where

$P^t = \{P_1 \ P_2 \ \ldots \ P_m\}$

= a set of m *model constants* which appear in the system's differential equations.

In many applications, the initial conditions $X(t_o)$ of (4.110) will be poorly known, as well as one or more elements of the model parameter vector P. Thus it may be necessary to estimate both $X(t_o)$ and P based upon measurements of X(t) or a function thereof. As will be seen in the applications of Chapter 6, conventional estimation processes will require the partial derivative matrices

$$[\Phi(t,t_o)] = \left[\frac{\partial X(t)}{\partial X(t_o)}\right] \tag{4.111}$$

and

$$[\Psi(t,t_o)] = \left[\frac{\partial X(t)}{\partial P}\right] \quad . \tag{4.112}$$

We now investigate methods for calculating these for derivative matrices.

Equation (4.110) can be written in integral form as

$$X(t) = X(t_o) + \int_{t_o}^{t} F(\tau,X,P)d\tau \tag{4.113}$$

from which it follows that

$$[\Phi(t\ t_o)] \equiv \left[\frac{\partial X(t)}{\partial X(t_o)}\right] = \{I\} + \int_{t_o}^{t} \left[\frac{\partial F}{\partial x(\tau)}\right] \left[\frac{\partial x(\tau)}{\partial x(t_o)}\right] d\tau \tag{4.114}$$

and

$$[\Psi(t,t_o)] \equiv \left[\frac{\partial X(t)}{\partial P}\right] = \int_{t_o}^{t} \left(\left[\frac{\partial F}{\partial P}\right] + \left[\frac{\partial F}{\partial x(\tau)}\right] \left[\frac{\partial x(\tau)}{\partial P}\right]\right) d\tau \tag{4.115}$$

Taking the time derivative of (4.114) and (4.115), it follows that the desired derivative matrices satisfy the first order linear differential equations

$$\frac{d}{dt}\{\Phi(t,t_o)\} = \{A(t)\}\{\Phi(t,t_o)\},\ \{\Phi(t_o,t_o)\} = \{I\} \tag{4.116}$$

and

$$\frac{d}{dt}\{\Psi(t,t_o)\} = \{A(t)\}\{\Psi(t,t_o)\} + \left[\frac{\partial F(t,X,P)}{\partial P}\right],\ \{\Psi(t_o,t_o)\}=\{0\} \tag{4.117}$$

where

$$\{A(t)\} \equiv \left[\frac{\partial F(t,X,P)}{\partial x(t)}\right] \quad . \tag{4.118}$$

Observe that $\{A(t)\}$ and $\{\frac{\partial F(t,X,P)}{\partial P}\}$ in eqns. (4.116) and (4.117) depend upon $X(t)$. If numerical methods are required to solve the differential equations (4.110) for $X(t)$, it is usually convenient to employ the same numerical process to simultaneously integrate (4.116) and (4.117) to obtain $\{\Phi(t,t_o)\}$ and $\{\Psi(t,t_o)\}$. Clearly, if the original system can be solved analytically for $X(t)$, then the partial derivatives can be taken formally and analytical solutions determined for $\{\Phi(t,t_o)\}$ and $\{\Psi(t,t_o)\}$.

As is evident by comparison of (4.116) and (4.10), the derivative matrix has the interpretation

$$\{\delta X(t)\} = \{\Phi(t,t_o)\}\ \{\delta X(t_o)\}$$

where $\{\delta X\}$ are small variations about a reference solution of (4.110). One important conclusion of the above is that if the original nonlinear system can be solved analytically, then the linear variational equations (4.10), (4.116) and (4.117) can also be solved analytically (i.e., their solution is reduced to a process of formal partial differentiation).

Example 4.10 Analytical Partial Derivatives of the Elliptic Two Body Problem

Referring to Example 4.6, the equations of motion (4.95) can be written as the equivalent first order system

$$\begin{aligned}
\dot{x}_1 &= x_4 \\
\dot{x}_2 &= x_5 \\
\dot{x}_3 &= x_6 \\
\dot{x}_4 &= -\mu\{x_1^2 + x_2^2 + x_3^2\}^{-3/2}\ x_1 \qquad (4.119) \\
\dot{x}_5 &= -\mu\{x_1^2 + x_2^2 + x_3^2\}^{-3/2}\ x_2 \\
\dot{x}_6 &= -\mu\{x_1^2 + x_2^2 + x_3^2\}^{-3/2}\ x_3
\end{aligned}$$

Considering $P = \mu$, the matrices in the differential equations (4.116) and (4.117) are explicitly

$$\{\Phi(t,t_o)\} = \left[\begin{array}{ccc:ccc} \frac{\partial x_1(t)}{\partial x_1(t_o)} & \frac{\partial x_1(t)}{\partial x_2(t_o)} & \frac{\partial x_1(t)}{\partial x_3(t_o)} & \frac{\partial x_1(t)}{\partial x_4(t_o)} & \frac{\partial x_1(t)}{\partial x_5(t_o)} & \frac{\partial x_1(t)}{\partial x_6(t_o)} \\ \frac{\partial x_2(t)}{\partial x_1(t_o)} & \frac{\partial x_2(t)}{\partial x_2(t_o)} & \frac{\partial x_2(t)}{\partial x_3(t_o)} & \frac{\partial x_2(t)}{\partial x_4(t_o)} & \frac{\partial x_2(t)}{\partial x_5(t_o)} & \frac{\partial x_2(t)}{\partial x_6(t_o)} \\ \frac{\partial x_3(t)}{\partial x_1(t_o)} & \frac{\partial x_3(t)}{\partial x_2(t_o)} & \frac{\partial x_3(t)}{\partial x_3(t_o)} & \frac{\partial x_3(t)}{\partial x_4(t_o)} & \frac{\partial x_3(t)}{\partial x_5(t_o)} & \frac{\partial x_3(t)}{\partial x_6(t_o)} \\ \hdashline \frac{\partial x_4(t)}{\partial x_1(t_o)} & \frac{\partial x_4(t)}{\partial x_2(t_o)} & \frac{\partial x_4(t)}{\partial x_3(t_o)} & \frac{\partial x_4(t)}{\partial x_4(t_o)} & \frac{\partial x_4(t)}{\partial x_5(t_o)} & \frac{\partial x_4(t)}{\partial x_6(t_o)} \\ \frac{\partial x_5(t)}{\partial x_1(t_o)} & \frac{\partial x_5(t)}{\partial x_2(t_o)} & \frac{\partial x_5(t)}{\partial x_3(t_o)} & \frac{\partial x_5(t)}{\partial x_4(t_o)} & \frac{\partial x_5(t)}{\partial x_5(t_o)} & \frac{\partial x_5(t)}{\partial x_6(t_o)} \\ \frac{\partial x_6(t)}{\partial x_1(t_o)} & \frac{\partial x_6(t)}{\partial x_2(t_o)} & \frac{\partial x_6(t)}{\partial x_3(t_o)} & \frac{\partial x_6(t)}{\partial x_4(t_o)} & \frac{\partial x_6(t)}{\partial x_5(t_o)} & \frac{\partial x_6(t)}{\partial x_6(t_o)} \end{array}\right] \tag{4.120}$$

$$\{\Psi(t,t_o)\} = \left\{\begin{array}{c} \frac{\partial x_1(t)}{\partial \mu} \\ \frac{\partial x_2(t)}{\partial \mu} \\ \frac{\partial x_3(t)}{\partial \mu} \\ \frac{\partial x_4(t)}{\partial \mu} \\ \frac{\partial x_5(t)}{\partial \mu} \\ \frac{\partial x_6(t)}{\partial \mu} \end{array}\right\} \tag{4.121}$$

$$r^2 \equiv x_1^2(t) + x_2^2(t) + x_3^2(t) \tag{4.122}$$

$$\{A(t)\} \equiv \{\frac{\partial F}{\partial x}\} =$$

$$\left[\begin{array}{ccc:ccc} 0 & 0 & 0 & 1 & 0 & 0 \\ 0 & 0 & 0 & 0 & 1 & 0 \\ 0 & 0 & 0 & 0 & 0 & 1 \\ \hdashline \frac{3\mu x_1^2}{r^5} - \frac{\mu}{r^3} & \frac{3\mu x_1 x_2}{r^5} & \frac{3\mu x_1 x_3}{r^5} & 0 & 0 & 0 \\ \frac{3\mu x_1 x_2}{r^5} & \frac{3\mu x_2^2}{r^5} - \frac{\mu}{r^3} & \frac{3\mu x_2 x_3}{r^5} & 0 & 0 & 0 \\ \frac{3\mu x_1 x_3}{r^5} & \frac{3\mu x_2 x_3}{r^5} & \frac{3\mu x_3^2}{r^5} - \frac{\mu}{r^3} & 0 & 0 & 0 \end{array}\right] \tag{4.123}$$

$$\frac{\partial F}{\partial P} \equiv \frac{\partial F}{\partial \mu} = \left\{\begin{array}{c} 0 \\ 0 \\ 0 \\ \cdots\cdots \\ -x_1/r^3 \\ -x_2/r^3 \\ -x_3/r^3 \end{array}\right\} \tag{4.124}$$

The "brute force" approach to determination of $\{\Phi(t,t_o)\}$ and $\{\Psi(t,t_o)\}$ would be to attempt formal analytical or numerical solutions of the differential equations (4.116) and (4.117). However, we can make efficient use of the fact that the analytical solution is available for X(t) (see Example 4.6) to determine the desired derivative solution by partial differentiation of the equations in-order-of solution {leading to equations (4.96)}. The appropriate equations in order of solution for the partial derivatives are:

$$\begin{Bmatrix} c_1 \\ c_2 \\ c_3 \\ c_4 \\ c_5 \\ c_6 \end{Bmatrix} \equiv \begin{Bmatrix} x(t_o) \\ y(t_o) \\ z(t_o) \\ \dot{x}(t_o) \\ \dot{y}(t_o) \\ \dot{z}(t_o) \end{Bmatrix} \tag{4.125}$$

then

$$\frac{\partial}{\partial c_k} \begin{Bmatrix} x \\ y \\ z \end{Bmatrix} = f \frac{\partial}{\partial c_k} \begin{Bmatrix} c_1 \\ c_2 \\ c_3 \end{Bmatrix} + \frac{\partial f}{\partial c_k} \begin{Bmatrix} c_1 \\ c_2 \\ c_3 \end{Bmatrix} + g \frac{\partial}{\partial c_k} \begin{Bmatrix} c_4 \\ c_5 \\ c_6 \end{Bmatrix} + \frac{\partial g}{\partial c_k} \begin{Bmatrix} c_4 \\ c_5 \\ c_6 \end{Bmatrix} \quad k = 1,2,\ldots,6 \tag{4.126}$$

where

$$\frac{\partial}{c_1} \begin{Bmatrix} c_1 \\ c_2 \\ c_3 \end{Bmatrix} = \begin{Bmatrix} 1 \\ 0 \\ 0 \end{Bmatrix} , \quad \frac{\partial}{\partial c_2} \begin{Bmatrix} c_1 \\ c_2 \\ c_3 \end{Bmatrix} = \begin{Bmatrix} 0 \\ 1 \\ 0 \end{Bmatrix} , \text{ etc.}$$

$$\frac{\partial f}{\partial c_k} = \frac{-a}{r_o} \sin\phi \frac{\partial \phi}{\partial c_k} - \frac{(1-\cos\phi)}{r_o^2}\left(r_o \frac{\partial a}{\partial c_k} - a \frac{\partial r_o}{\partial c_k} \right) \tag{4.127}$$

$$\frac{\partial g}{\partial c_k} = \frac{-a^{3/2}(1-\cos\phi)}{\sqrt{\mu}} \frac{\partial \phi}{\partial c_k} - \frac{3\sqrt{a}(\phi-\sin\phi)}{2\sqrt{\mu}} \frac{\partial a}{\partial c_k} \tag{4.128}$$

$$\frac{\partial r_o}{\partial c_k} = \frac{c_k}{r_o} , \quad k = 1,2,3 \tag{4.129}$$

$$\frac{\partial r_o}{\partial c_k} = 0 , \quad k = 4,5,6 \tag{4.130}$$

$$\frac{\partial a}{\partial c_k} = \frac{2c_k}{a^2 r_o^3} , \quad k = 1,2,3 \tag{4.131}$$

$$\frac{\partial a}{\partial c_k} = \frac{2c_k}{\mu a^2} , \quad k = 4,5,6 \tag{4.132}$$

$$\frac{\partial\phi}{\partial c_k} = \left[\sin\phi \frac{\partial c_o}{\partial c_k} - (1-\cos\phi) \frac{\partial}{\partial c_k} \frac{D_o}{\sqrt{\mu a}}\right]\left[1-c_o\cos\phi + \frac{D_o}{\sqrt{\mu a}} \sin\phi\right]^{-1}$$

$$- \frac{3}{2a}\{\phi - c_o\sin\phi + \frac{D_o}{\sqrt{\mu a}} (1 - \cos\phi)\}\{1 - c_o \cos\phi + \frac{D_o}{\sqrt{\mu a}}\sin\phi\}^{-1} \tag{4.133}$$

$$\frac{\partial c_o}{\partial c_k} = \frac{r_o}{a^2} \frac{\partial a}{\partial c_k} - \frac{1}{a} \frac{\partial r_o}{\partial c_k} \tag{4.174}$$

$$\frac{\partial}{\partial c_k}\left(\frac{D_o}{\sqrt{\mu a}}\right) = \frac{1}{\sqrt{\mu a}} \frac{\partial D_c}{\partial c_k} - \frac{1}{\sqrt{\mu}} \frac{D_o}{2a3/2} \frac{\partial a}{\partial c_k} \tag{4.135}$$

$$\frac{\partial D_o}{\partial c_1} = c_4 , \frac{\partial D_o}{\partial c_2} = c_5 , \frac{\partial D_o}{\partial c_3} = c_6$$

$$\frac{\partial D_o}{\partial c_4} = c_1 , \frac{\partial D_o}{\partial c_5} = c_2 , \frac{\partial D_o}{\partial c_6} = c_3 \tag{4.136}$$

4.6 Remarks

The essence-oriented discussion of differential equations of the present chapter, while adequate background for following the discussion of chapters 5 and 6, will likely prove incomplete in many applications. In particular, the terse discussion of numerical methods is §4.4 was limited to *single step methods* {methods which require only a single state $X(t_k)$ to approximate $X(t_{k+1})$}. Numerous practical problems are more efficiently solved by *multi-step methods* {procedures which employ a number of previous state approximations to extrapolate to $X(t_{k+1})$; the number of back states required defines the order of the method}. A treatment of multi-step methods here would have necessitated treatment of the calculus of finite differences; development of this material was judged outside the scope of the present monograph. The interested reader is referred to references 4.7 and 4.8 for excellent discussions and comparison of modern multi-step methods.

Also, conspicous by its lack of coverage here is perturbation theory; references 4.4, 4.5, and 4.6 document perturbation methods which are exceptionally valuable tools for solving weakly non-linear differential equations.

The results of the present chapter does provide an adequate basis for solving the differential equations encountered in a substantial fraction of practical applications, and provides a foundation for further study.

4.7 *Exercises for Chapter 4*

4.1 Given the coupled nonlinear second order system

$$\ddot{x} = -x + axy + u_x(t)$$

$$\ddot{y} = -y + bxy + u_y(t)$$

with a,b constants and $u_x(t)$, $u_y(t)$ specified forcing functions.

Rearrange the above eqns. to the form of eqns. (4.1). Determine the associated linear differential equations whose solutions yield the derivative matrices

$$\{\Phi(t,t_o)\} = \begin{bmatrix} \frac{\partial x(t)}{\partial x(t_o)} & \frac{\partial x(t)}{\partial y(t_o)} & \frac{\partial x(t)}{\partial \dot{x}(t_o)} & \frac{\partial x(t)}{\partial \dot{y}(t_o)} \\ \frac{\partial y(t)}{\partial x(t_o)} & \frac{\partial y(t)}{\partial y(t_o)} & \frac{\partial y(t)}{\partial \dot{x}(t_o)} & \frac{\partial y(t)}{\partial \dot{y}(t_o)} \\ \frac{\partial \dot{x}(t)}{\partial x(t_o)} & \frac{\partial \dot{x}(t)}{\partial y(t_o)} & \frac{\partial \dot{x}(t)}{\partial \dot{x}(t_o)} & \frac{\partial \dot{x}(t)}{\partial \dot{y}(t_o)} \\ \frac{\partial \dot{y}(t)}{\partial x(t_o)} & \frac{\partial \dot{y}(t)}{\partial y(t_o)} & \frac{\partial \dot{y}(t)}{\partial \dot{x}(t_o)} & \frac{\partial \dot{y}(t)}{\partial \dot{y}(t_o)} \end{bmatrix}$$

$$\{\Psi(t,t_o)\} = \begin{bmatrix} \frac{\partial x(t)}{\partial a} & \frac{\partial x(t)}{\partial b} \\ \frac{\partial y(t)}{\partial a} & \frac{\partial y(t)}{\partial b} \\ \frac{\partial \dot{x}(t)}{\partial a} & \frac{\partial \dot{x}(t)}{\partial b} \\ \frac{\partial \dot{y}(t)}{\partial a} & \frac{\partial \dot{y}(t)}{\partial b} \end{bmatrix} .$$

4.2 A *symplectic matrix* is a 2n x 2n matrix with the defining property that

$$A^t\ J\ A = J$$

where J is a matrix analogy of the scalar complex number $j^2 = -1$; J is defined as the 2n x 2n matrix

$$J = \begin{bmatrix} 0 & I_n \\ -I_n & 0 \end{bmatrix}, \quad J \cdot J = -I_{2n}$$

An important consequence of the symplectic property is that the inverse can be obtained by the simple rearrangement of A's elements as

$$A^{-1} = -JA^tJ = \begin{bmatrix} A_{22}^t & -A_{12}^t \\ -A_{21} & A_{11}^t \end{bmatrix}$$

where A is partitioned into n x n sub-matrices as

$$A = \begin{bmatrix} A_{11} & A_{12} \\ A_{21} & A_{22} \end{bmatrix}$$

This non-numerical inversion is a most important computational advantage that sympletic matrices have in common with orthogonal matrices.

The 6 x 6 two-body state transition matrix $\{\Phi(t,t_o)\}$ satisfies

$$\frac{d}{dt}\{\Phi(t,t_o)\} = \overset{\begin{matrix}3x3 & 3x3\end{matrix}}{\begin{bmatrix} 0 & I \\ G & 0 \end{bmatrix}} \{\Phi(t,t_o)\}$$

where, from eqn. (4.123)

$$G = \text{"gravity gradient"} = \frac{3\mu}{r^5}\begin{bmatrix} (x^2-r^2/3) & xy & xz \\ xy & (y^2-r^2/3) & yz \\ xz & yz & (z^2-r^2/3) \end{bmatrix}$$

Prove that $\Phi(t,t_o)$ is symplectic.

4.3 Write a computer program to solve for $x(t)$, $y(t)$, $\dot{x}(t)$, $\dot{y}(t)$ for $0 \leq t \leq 10$ in example (4.1), using $a = b = 1$ and the initial conditions

$$x(t_o) = y(t_o) = 1.0$$

$$\dot{x}(t_o) = \dot{y}(t_o) = 0.0$$

Use the fourth order Runge-Kutta algorithm developed in §4.4.3. Vary h and plot relative closure error versus h as in figure 4.1; determine the maximum h consistent with $\Delta = 10^{-5}$ {use $K_i = 1.0$, $i = 1,2,2,4$ in equation (4.109)}.

4.4 Program the analytical solution for the elliptic two body problem governed by

$$\begin{aligned} \ddot{x} &= -\frac{\mu}{r^3} x \\ \ddot{y} &= -\frac{\mu}{r^3} y \\ \ddot{z} &= -\frac{\mu}{r^3} z \end{aligned} \tag{4.95}$$

as given in example 4.6, compute the state history $x(t)$, $y(t)$, $z(t)$ at an interval of 200 seconds for 5000 seconds. The initial conditions are given as

$$\begin{aligned} x(t_o) &= 7000 \text{ km} \\ y(t_o) &= 10 \text{ km} \\ z(t_o) &= 20 \text{ km} \\ \dot{x}(t_o) &= 4 \text{ km/sec} \\ \dot{y}(t_o) &= 7 \text{ km/sec} \\ \dot{z}(t_o) &= 2 \text{ km/sec} \end{aligned}$$

Also, solve the differential equations (4.95) using fourth order Runge-Kutta and compare the numerical solution with the analytical solution for h = 200, 100, 50, 25, 12.5, and 6.25 second step sizes. Compare the *absolute* and *relative* closure error norms and construct plots for each analogous to figure 4.1.

4.8 References for Chapter 4

4.1 Herrick, S., Astrodynamics Vol II, Van Nostrand Reinhold (1971) ch 7.

4.2 Ince, E.L., Ordinary Differential Equations, Longmans, London (1926), pp 540-547.

4.3 Ogata, K., State Space Analysis for Engineers, Prentice Hall (1967), pp 120-173.

4.4 Meirovitch, L., Methods of Analytical Dynamics, McGraw-Hill (1970), ch 5, 6, and 8.

4.5 Nayfeh, A., Perturbation Methods, Wiley (1973), ch 1-5.

4.6 Cole, J.D., Perturbation Methods in Applied Mathematics, Wiley (1968), ch 1-4.

4.7 Milne, W.E., Numerical Solution of Differential Equations, Dover (1970), pp 34-36.

4.8 Merson, R.H., "Numerical Integration of Differential Equations...", contract report ESOC 377/71/HR, published by the Royal Aircraft Establishment, Farnborough, Hants, Great Britain, May 1973.

5

Estimation of Dynamical Systems

5.1 *Preliminary Remarks*

In the developments of the first three chapters, estimation concepts are formulated and applied to systems whose measured variables are related to the estimated parameters by *algebraic* equations. The present chapter extends these results to allow estimation of parameters embedded in the model of a *dynamical system*, the model usually includes both *algebraic* and *differential* equations.

We will find that the batch estimation results of §1.2, 1.3 and 1.5, developed for estimation of *algebraic* systems, remain valid for estimation of *dynamical* systems; upon making the appropriate new interpretations of the matrices involved in the estimation algorithms. In the event that the differential equations have explicit algebraic solutions, of course, the entire model becomes algebraic equations and the methods of the first three chapters apply immediately. On the other hand, we'll find that the sequential estimation results of §1.4 must be extended to properly account for "motion" of the dynamical system between measurement and estimation epochs.

The formulations of the present chapter are developed as natural extensions of the estimation methods of the first three chapters using differential equation models and notations of chapter 4. In chapter 6, several prototype applications are summarized and several results from the recent literature are reviewed.

5.2 Initial State Estimation

A frequently occurring estimation problem has the following structure:

The state is described by a system of (generally nonlinear) differential equations

$$\dot{X} = F(t, X) \tag{5.1}$$

Discrete measurements of Y(t) are available, where the measured quantities are functions of the state variables

$$\tilde{Y}(t_j) = H(X(t_j)) + V(t_j); \quad j = 0, 1, 2 \ldots, M. \tag{5.2}$$

where

X = an nx1	vector of state variables
Y = an mx1	vector of measurable variables
$F(t,X)$ = an nx1	vector of generally nonlinear functions
$H(X)$ = an mx1	vector of generally nonlinear functions
$V(t_j)$ = an mx1	vector of measurement errors with the properties $E\{V(t_j)\} = 0$, $E\{V(t_\ell)V^t(t_m)\} = R_{\ell m}$.

Given the model functions F and H and the M + 1 sets of measurements $\tilde{Y}(t_k)$, it is desired to determine optimal estimates of the initial state vector $X(t_o)$.

An excellent example problem having this structure is the spacecraft orbit determination problem {the measurements (5.2) would likely be the angles, angular rates, distance, distance rate-of-change, etc. with respect to one or more tracking stations, the differential equations (5.1) would be of the form of equations (4.5) or (4.85)}.

Thus the mathematical model relating the measurements $\tilde{Y}$ to the to-be-estimated-parameter vector $X(t_o)$ involves the system of differential equations (5.1). Equations (5.1) could be written in integral equation form as

$$X(t_j) = X(t_o) + \int_{t_o}^{t_j} F(\tau, X(\tau))\, d\tau \tag{5.3}$$

Then the equation

$$\tilde{Y}(t_j) = \text{Function } (t_j, X(t_o)) + V(t_j) \tag{5.4}$$

implies eqn. (5.3) substituted into (5.2). Eqns. (5.4) have the structure of the models considered in §1.8; therefore the Gaussian least squares differential algorithm

$$\hat{X}(t_o)^{(k+1)} = \hat{X}(t_o)^{(k)} + \{(A^t WA)^{-1} A^t W\, \Delta Y\}^{(k)} \tag{5.5}$$

will provide successive approximations for the initial condition estimates, where

$$A \equiv \begin{bmatrix} \left.\dfrac{\partial Y(t_o)}{\partial X(t_o)}\right|_k \\ \hline \left.\dfrac{\partial Y(t_1)}{\partial X(t_o)}\right|_k \\ \hline \vdots \\ \hline \left.\dfrac{\partial Y(t_M)}{\partial X(t_o)}\right|_k \end{bmatrix} \tag{5.6}$$

$$\Delta Y = \begin{Bmatrix} \tilde{Y}(t_o) - Y(t_o)^{(k)} \\ \hline \tilde{Y}(t_1) - Y(t_1)^{(k)} \\ \hline \vdots \\ \hline \tilde{Y}(t_M) - Y(t_M)^{(k)} \end{Bmatrix} \tag{5.7}$$

$$X(t_j)^{(k)} = \hat{X}(t_o)^{(k)} + \int_{t_o}^{t_j} F(\tau, X(\tau))d\tau \tag{5.8}$$

$= $ computed value of X at time t_j based upon estimate $\hat{X}(t_o)^{(k)}$ of $X(t_o)$

$$Y(t_j)^{(k)} = H(X(t_j)^{(k)}) \tag{5.9}$$

$$W = E\{V_j V_j^t\}^{-1} \tag{5.10}$$

$$V_j^t \equiv \{V^t(t_o) \vdots V^t(t_1) \vdots \ldots \vdots V^t(t_M)\} \tag{5.11}$$

The submatrices of derivatives in the A matrix (5.6) are determined using the chain rules and methods of §4.5 as follows: From equations (5.2)

$$\left[\frac{\partial Y(t_j)}{\partial X(t_o)}\right] = \left[\frac{\partial H(X(t_j))}{\partial X(t_j)}\right] \left[\frac{\partial X(t_j)}{\partial X(t_o)}\right] \tag{5.12}$$

where the *state transition matrix* is determined by solving the system of linear differential equations

$$\frac{d}{dt}\left[\frac{\partial X(t)}{\partial X(t_o)}\right] = \left[\frac{\partial F(t,X(t))}{\partial X(t)}\right] \left[\frac{\partial X(t)}{\partial X(t_o)}\right] . \tag{5.13}$$

using the identity matrix as initial conditions.

The above process is summarized in Figure 5.1; it finds a wide family of applications and is embedded in many more involved dynamical estimation processes. Typical example applications are given in chapter 6.

5.3 Initial State and Model Parameter Estimation

The above formulation can be generalized to allow estimation not only of the initial state but, also, parameters which appear

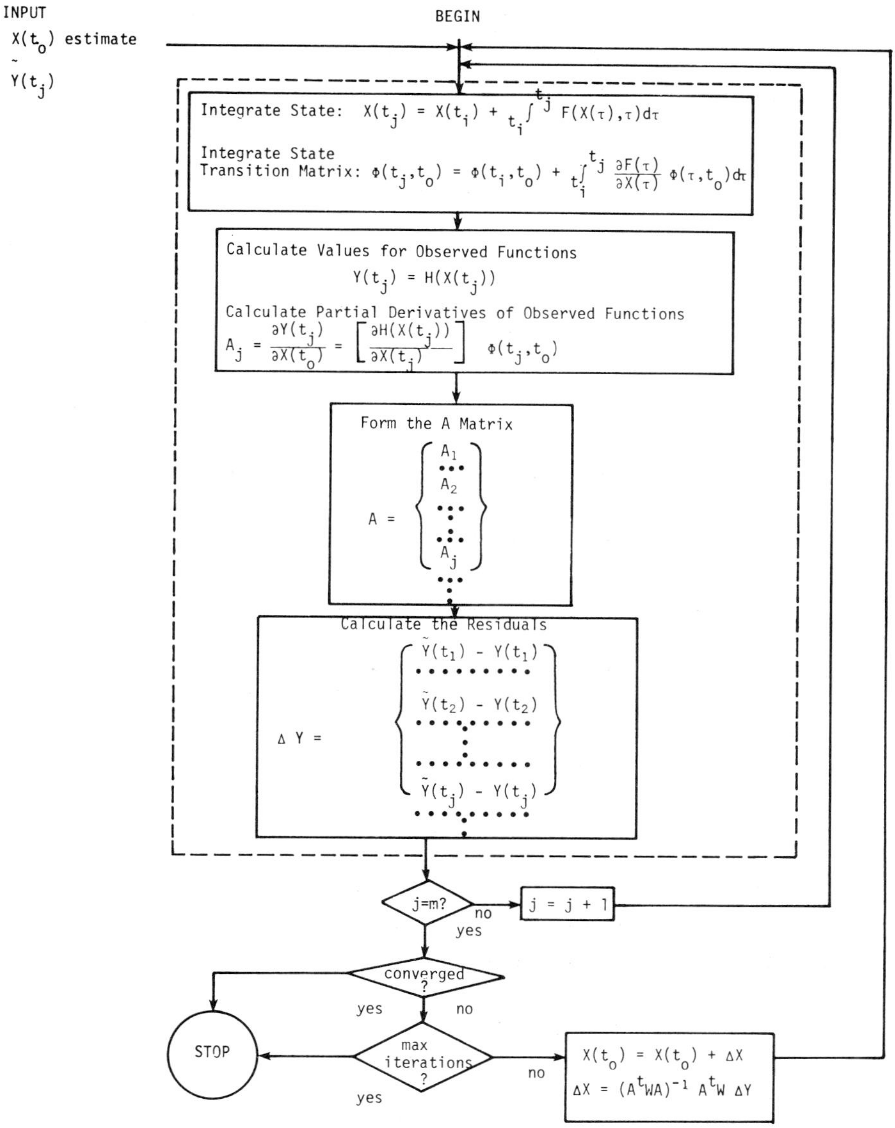

Figure 5.1 Initial State Estimation for Nonlinear Dynamical Systems Using Gaussian Least Square Differential Correction

in the state differential equations and/or, in the measurement model equations we state the problem as follows: the state satisfies the system of differential equations

$$\dot{X} = F(t, X, C) \tag{5.14}$$

and the measurements are modeled by

$$\tilde{Y} = H(X,K) + V \tag{5.15}$$

where C and K are vectors of unknown parameters that appear in the system dynamic model and measurement model, respectively.

$$\mathbf{X} = \left\{\begin{array}{c} X(t_o) \\ \hline C \\ \hline K \end{array}\right\} , \tag{5.16}$$

a vector containing all three sets of parameters to-be-estimated. The differential correction algorithm to determine the successive approximations is now applied to estimating $\mathbf{X}$ as

$$\hat{\mathbf{X}}^{(k+1)} = \hat{\mathbf{X}}^{(k)} + [(\mathbf{A}^t\ W\ \mathbf{A})^{-1}\ \mathbf{A}^t\ W\ \Delta Y]^{(k)} \tag{5.17}$$

where

$$\mathbf{A} = \left[\frac{\partial Y}{\partial X(t_o)} \ \vdots \ \frac{\partial Y}{\partial C} \ \vdots \ \frac{\partial Y}{\partial K} \right] \tag{5.18}$$

The first partition of $\mathbf{A}$ is recognized as the A matrix of eqn. (5.6) and is calculated in the identical fashion. The second partition is determined as follows:

From §4.5, $\left[\frac{\partial X(t)}{\partial C}\right]$ is the solution of the differential equation

$$\frac{d}{dt}\left[\frac{\partial X}{\partial C}\right] = \left[\frac{\partial F}{\partial C}\right] + \left[\frac{\partial F}{\partial X}\right]\left[\frac{\partial X}{\partial C}\right] \tag{5.19}$$

with the zero matrix as initial condition. It then follows from the chain rule from (5.15) that

$$\left[\frac{\partial Y}{\partial C}\right] = \left[\frac{\partial H}{\partial X}\right]\left[\frac{\partial X}{\partial C}\right] . \tag{5.20}$$

The final partition of (5.18) does not involve variables or parameters embedded in differential equations, these partials may be determined by direct differentiation of equations (5.15) as

$$\left[\frac{\partial Y}{\partial K}\right] = \left[\frac{\partial H}{\partial K}\right] \tag{5.21}$$

Perhaps the above ideas are best illustrated by a simplified example, such as the one immediately following.

Example 5.1 Satellite Tracking Via Laser Ranging

A satellite S is moving about the Earth's center C and is being tracked by a laser beam fired from point T on the Earth's surface. For simplicity, the Earth's rotation is ignored. The instrumentation is capable of measuring the elapsed time from firing the laser until return light (reflected off "corner reflectors" on the satellite) is detected. Since the speed of light is known, then this provides a measure of the *instantaneous* (to the extent that the satellite velocity of about 8 km/sec is negligible to the speed of light $\simeq 3\text{x}10^8$ km/sec) distance or range from the tracking station T to the satellite S.

For simplicity, let the state history evolve as the solution of the 2-body equations of motion

$$\begin{aligned}
\dot{x}_1 &= x_4 \\
\dot{x}_2 &= x_5 \\
\dot{x}_3 &= x_6 \\
\dot{x}_4 &= -GM\ x_1\{x_1^2 + x_2^2 + x_3^2\}^{-3/2} \\
\dot{x}_5 &= -GM\ x_2\{x_1^2 + x_2^2 + x_3^2\}^{-3/2} \\
\dot{x}_6 &= -GM\ x_3\{x_1^2 + x_2^2 + x_3^2\}^{-3/2}
\end{aligned} \tag{5.22}$$

The spacecraft S coordinates relative to the Earth center C are (x_1, x_2, x_3), and the coordinates of the tracking station relative to C are denoted as (x_τ, y_τ, z_τ). The measured range distance (being measured by the laser beam round trip times) is clearly modeled by

$$\tilde{d} = \{(x_1 - x_\tau)^2 + (x_2 - y_\tau)^2 + (x_3 - z_\tau)^2\}^{\frac{1}{2}} + v \tag{5.23}$$

Suppose that it was desired to improve not only available estimates for the orbit initial conditions $\{x_i(t_o)\}$, but also the location (x_τ, y_τ, z_τ) of the laser tracking station; given nine or more measurements of (5.23), at known measurement times.

Equation (5.23) is of the same form as equation (5.15) with the K vector identified as

$$K = \begin{Bmatrix} x_\tau \\ y_\tau \\ z_\tau \end{Bmatrix} .$$

Equations (5.22) are of the same form as equations (5.14), but unless we consider the gravitational-mass constant GM uncertain (it currently has eight significant digits) then the C-vector is void.

Thus the successive approximations are determined according to eqn. (5.17) with

$$\mathbf{X} = \begin{Bmatrix} x_1(t_o) \\ \vdots \\ x_6(t_o) \\ \hline x_\tau \\ y_\tau \\ z_\tau \end{Bmatrix}$$

$$A \equiv \left[\frac{\partial Y}{\partial X(t_o)} \ \vdots \ \frac{\partial Y}{\partial K} \right]$$

and

$$\left[\frac{\partial Y}{\partial K} \right] \equiv \begin{bmatrix} \frac{\partial d(t_o)}{\partial x_\tau} & \frac{\partial d(t_o)}{\partial y_\tau} & \frac{\partial d(t_o)}{\partial z_\tau} \\ & \vdots & \\ \frac{\partial d(t_m)}{\partial x_\tau} & \frac{\partial d(t_m)}{\partial y_\tau} & \frac{\partial d(t_m)}{\partial z_\tau} \end{bmatrix}$$

The state variables and state transition matrix, for this case, can be determined analytically as shown in example 4.10.

The differential correction process of the current discussion and §5.2 can easily be modified to account for uncertainty estimates associated with the apriori estimates. From the developments of §2.5, we can replace equation (5.17) by

$$\hat{X}^{(k+1)} = \hat{X}^{(k)} + \left[(A^t W A + \Lambda^{-1}_{xx_a})^{-1} (A^t W \Delta Y + \Lambda^{-1}_{xx_a} \hat{X}) \right]^{(k)} \tag{5.23}$$

where

Λ_{xx_a} = the covariance matrix associated with the errors in the starting estimates $\hat{X}^{(o)}$.

5.4 *Sequential State Estimation for Linear Dynamic Systems*

In the estimation formulations developed previously, it has been assumed that a specific set of parameters are being estimated; additional data were allowed, but the parameters being estimated remained unchanged. A more complicated situation arises whenever the set of parameters being estimated is allowed to change during the estimation process. To motivate the discussion, consider real time estimation of the state of a manuevering spacecraft. As each subset of observations becomes available, it is desired to obtain an optimal estimate of the state *at that instant* in order to, for example, provide the best current information to base control decisions upon.

In the accepted estimation and control theory terminology; if the *current* state is being estimated, the estimation process is referred to as *filtering*; if the state at a *previous* time is being estimated, then the estimation process is referred to as *smoothing*; if a *future* state is being estimated, the process is referred to as *prediction*. While filtering, smoothing, and prediction are obviously related, we will find that it is possible to develop the "machinery" applicable to all three by considering in detail only the filtering problem. The sequential estimation results of Chapter 1 will be combined with the linear differential equation results developed in Chapter 4; considering unforced (homogeneous) dynamic systems first and then forced dynamic systems.

5.4.1 Unforced Linear Systems

Assume that the state of a system of interest is governed by the $n_x \times 1$ set of ordinary differential equations

$$\dot{X}(t) = A(t)\,X(t) \tag{5.24}$$

It is occasionally possible to directly measure the state variables resulting from solution of (5.24); these cases are in the minority of those encountered in practice. {For example, it is generally impossible to directly observe the departures of a spacecraft from a nominal trajectory at the corresponding time!} The observable quantities $Y^t = (y_1 \ldots y_m)$ are generally non-linear functions of the system state variables {e.g., a radar might observe range, range-rate, azimuth, and elevation}. Consider, for the moment, that the observations are linear combinations of the state variables as

$$\tilde{Y}(t) = H(t)\,X(t) + V(t)\ , \tag{5.25}$$

where H(t) is an m x n_x known matrix, and V(t) is an m x 1 matrix of measurement errors with zero mean and known covariance $\Lambda_{VV} \equiv E(VV^t)$.

In this section, we are concerned with the case in which observations become available at discrete times $\{t_1, t_2, \ldots, t_k, t_{k+1}\}$. The assumption is made that the corresponding measurement errors V(1),V(2),...,V(k),V(k+1) in (5.25) are a "white noise" random disturbance; which means that the errors are not correlated foreward or backward in time so that

$$E\{V(i)V^t(j)\} = \begin{cases} 0 & i \neq j \\ \Lambda_{v_i v_i} & i = j \end{cases} \tag{5.26}$$

This requirement preserves the block diagonal structure of the covariance and weight matrices.

The sequential filtering problem can now be stated as follows: assume that k subsets of data*

$$\begin{aligned} \tilde{Y}(1) &= H(1)X(1) + V(1) \\ \tilde{Y}(2) &= H(2)X(2) + V(2) \\ &\vdots \\ \tilde{Y}(k) &= H(k)X(k) + V(k) \end{aligned} \tag{5.27}$$

have been collected. The state at the k times corresponding to the observation times of (5.25) can be linearly related via transition matrices to the state at some arbitrary time t_j as

$$\begin{aligned} X(1) &= \Phi(1,j)X(j) \\ X(2) &= \Phi(2,j)X(j) \\ &\vdots \\ X(k) &= \Phi(k,j)X(j) \end{aligned} \tag{5.28}$$

---

*f(j) = shorthand for $f(t_j)$.

Substitution of (5.28) into (5.27) then yields the k observation subset equations

$$\begin{aligned} \tilde{Y}(1) &= H(1)\Phi(1,j)X(j) + V(1) \\ \tilde{Y}(2) &= H(2)\Phi(2,j)X(j) + V(2) \\ &\vdots \\ \tilde{Y}(k) &= H(k)\Phi(k,j)X(j) + V(k) \end{aligned} \tag{5.29}$$

Introducing the definitions

$$\tilde{\mathbf{Y}}_k = \begin{Bmatrix} \tilde{Y}(1) \\ \tilde{Y}(2) \\ \vdots \\ \tilde{Y}(k) \end{Bmatrix} \tag{5.30}$$

$$\mathbf{A}_k = \begin{bmatrix} H(1)\ \Phi(1,j) \\ H(2)\ \Phi(2,j) \\ \vdots \\ H(k)\ \Phi(k,j) \end{bmatrix} \tag{5.31}$$

$$\mathbf{V}_k = \begin{Bmatrix} V(1) \\ V(2) \\ \vdots \\ V(k) \end{Bmatrix} \tag{5.32}$$

Then "merged" observation equations can be written as

$$\tilde{\mathbf{Y}}_k = \mathbf{A}_k X(j) + \mathbf{V}_k \tag{5.33}$$

The optimal minimal variance estimate $\hat{X}_k(j)$ of X(j) *could* be computed from the normal equations (1.30), with the weighting matrix

$$\mathbf{W}_k \equiv \Lambda^{-1}_{v_k v_k} = \begin{bmatrix} \Lambda_{v_1 v_1} & & & 0 \\ & \Lambda_{v_2 v_2} & & \\ & & \ddots & \\ 0 & & & \Lambda_{v_k v_k} \end{bmatrix}^{-1} \tag{5.34}$$

A sequential estimation process is desired, therefore, we assume that

$$\hat{X}_k(j)$$

and (5.35)

$$P_k(j) = (A_k^t W_k A_k)^{-1}$$

are available either as a direct solution of (1.30), from the sequential solution developed below, or as apriori estimates.

Based upon the "new" observation subset

$$\tilde{Y}(k+1) = H(k+1)\Phi(k+1,j)X(j) + V(k+1) \ , \tag{5.36}$$

it is desired to update $\hat{X}_k(j)$ to obtain $\hat{X}_{k+1}(j)$ the optimal estimate of X at time t_j, based upon all k+1 measurement subsets.

The results of Chapter 1, equations (1.93) and (1.94) can be employed to obtain the following pair of equations for updating X(j) and P(j) as

$$\hat{X}_{k+1}(j) = \hat{X}_k(j) + P_k(j)A_{k+1}^t(j)\left\{\Lambda_{v_{k+1}v_{k+1}} + A_{k+1}(j)P_k(j)A_{k+1}^t(j)\right\}^{-1}$$

$$x \ \left\{\tilde{Y}(k+1) - A_{k+1}(j)\hat{X}_k(j)\right\} \tag{5.37}$$

and

$$P_{k+1}(j) = P_k(j) - P_k(j)A_{k+1}^t(j)\left\{\Lambda_{v_{k+1}v_{k+1}} + A_{k+1}(j)P_k(j)A_{k+1}^t(j)\right\}^{-1}$$

$$x \ A_{k+1}(j)P_k(j) \tag{5.38}$$

where

$$A_{k+1}(j) \triangleq H(k+1)\Phi(k+1,j) \ . \tag{5.39}$$

The sequential filtering machinery can be obtained directly from equations (5.37) and (5.38). Recall that for a filtering estimation process, it is necessary that X(k+1) be estimated at $t=t_{k+1}$ based upon all data collected up to and including time t_{k+1}. Letting $t_j = t_{k+1}$ in (5.37) and (5.38) is clearly implied; note that, however, this yields terms containing $\hat{X}_k(k+1)$ {interpreted as the "best estimates of X and P at time t_{k+1} based upon the first k subsets of data"}. The paradox can be alleviated (for linear systems) by simply propagating $\hat{X}_k(k)$ and $P_k(k)$ forward in time as

$$\hat{X}_k(k+1) = \Phi(k+1,k)\hat{X}_k(k) \tag{5.40}$$

and {using equation (2.70)}

$$P_k(k+1) = \Phi(k+1,k)P_k(k)\Phi^t(k+1,k). \tag{5.41}$$

Equations (5.37), (5.38), and (5.39), now become {upon substitution of (5.40) and (5.41)} the final form for the sequential Kalman filter for linear unforced dynamic systems:

$$\hat{X}_{k+1}(k+1) = \Phi(k+1,k)\hat{X}_k(k) + K(k+1)\{\tilde{Y}(k+1) - H(k+1)\Phi(k+1,k)\hat{X}_x(k)\} \tag{5.42a}$$

and

$$P_{k+1}(k+1) = \{I - K(k+1)H(k+1)\}\ P_k(k+1), \tag{5.42b}$$

where

$$P_k(k+1) = \Phi(k+1,k)P_k(k)\Phi^t(k+1,k) \tag{5.42c}$$

$$K(k+1) = P_k(k+1)H^t(k+1)\ \{\Lambda_{v_{k+1}v_{k+1}} + H(k+1)P_k(k+1)H^t(k+1)\}^{-1}. \tag{5.42d}$$

For prediction and smoothing, the filtered state (5.42a) and the associated covariance matrix can be propagated forward or backwards in time to any desired time via

$$\hat{X}_{k+1}(j) = \Phi(j,k+1)\hat{X}_{k+1}(k+1) \tag{5.43}$$

and

$$P_{k+1}(j) = \Phi(j,k+1)\ P_{k+1}(k+1)\ \Phi^t(j,k+1). \tag{5.44}$$

5.4.2 Process Noise Forced Dynamic Systems

Here, the homogeneous results are generalized to allow the possibility of forcing ("process") noise on the system of differential equations. Thus, we assume that the system's dynamics are described by

$$\dot{X}(t) = A(t)X(t) + B(t)W(t) \tag{5.45}$$

W(t) is assumed to be a random sequence with known covariance

$$E\left\{W(t)W^t(t)\right\} = T(t,t) \tag{5.46}$$

and it is assumed that

$$E\left\{W(t)V^t(\tau)\right\} = 0 \tag{5.47}$$

holds for all t and τ. The observation model (5.25) and (5.26) are still assumed valid. From equation (4.73), the solution of (5.45) can be written as the recursion formulas

$$X(j+1) = \Phi(j+1,j)X(j) + U(j) \tag{5.48}$$

where

$$U(j) = \int_{t_j}^{t_{j+1}} \Phi(t_{j+1},\tau)B(\tau)W(\tau)d\tau\ . \tag{5.49}$$

Successive substitution of (5.48) into itself yields

$$X(k) = \Phi(k,j)X(j) + \sum_{i=j}^{k-1} \Phi(k,i+1)U(i)$$

from which

$$X(j) = \Phi(j,k)X(k) - \sum_{i=j}^{k-1} \Phi(j,i+1)U(i) \tag{5.50}$$

If we now restrict W(t) to be *piecewise constant* and specify that its values at $t_1, t_2, \ldots, t_j, t_{j+1} \ldots$ form a white noise sequence; then the covariance of W is

$$T(t_i,t_j) \equiv E\left\{W(i)W^t(j)\right\} = \begin{cases} T(j) & i=j \\ 0 & i\neq j \end{cases} \tag{5.51}$$

the covariance of U(j) is then

$$Q(j) \triangleq E\left\{U(j)U^t(j)\right\} \tag{5.52}$$

which, upon substituting (5.49) and taking W(j) from under the integral signs, becomes

$$Q(j) = E\left\{ \left[\int_{t_j}^{t_{j+1}} \Phi(t_{j+1},\tau)B(\tau)d\tau \right] W(j)W^t(j)\left[\int_{t_j}^{t_{j+1}} B^t(\tau)\Phi^t(t_{j+1},\tau)d\tau \right] \right\} \tag{5.53}$$

or

$$Q(j) = \Psi(j)T(j)\Psi^t(j) \tag{5.54}$$

where

$$\Psi(j) \equiv \int_{t_j}^{t_{j+1}} \Phi(t_{j+1},\tau)B(\tau)d\tau \,. \tag{5.55}$$

It follows from direct substitution of (5.49) (for W(t) piecewise constant) and making use of (5.47), that

$$E\left\{ U(j)V^t(j) \right\} = 0. \tag{5.56}$$

The sequential estimation machinery can now be developed by substituting (5.50) into the observation equations (5.27) to produce k subsets of observation equations as

$$\tilde{Y}(1) = H(1) \quad \Phi(1,k)X(k) - \sum_{i=1}^{k-1} \Phi(1,i+1)U(i) \quad + V(1)$$

$$\tilde{Y}(2) = H(2) \quad \Phi(2,k)X(k) - \sum_{i=2}^{k-1} \Phi(2,i+1)U(i) \quad + V(2)$$

$$\vdots \qquad\qquad \vdots$$

$$\tilde{Y}(k-1) = H(k-1)\Phi(k-1,k)X(k) - \Phi(k-1,k)U(k-1) \quad + V(k-1)$$

$$\tilde{Y}(k) = H(k)X(k) + V(k)$$

These k subsets or observation equations can be merged and written in matrix form as

$$\tilde{\mathbf{Y}}_k = \mathbf{A}_k X(k) - \mathbf{M}_k \mathbf{U}_k + \mathbf{V}_k \,, \tag{5.58}$$

where $\tilde{Y}_k$, A_k and V_k are still given by (5.30), (5.31), and (5.32), and where

$$M_k \triangleq \begin{bmatrix} H(1)\Phi(1,2) & H(1)\Phi(1,3)\ldots & H(1)\Phi(1,k) \\ 0 & H(2)\Phi(2,3)\ldots & H(2)\Phi(2,k) \\ 0 & 0 \quad . & H(3)\Phi(3,k) \\ \vdots & \vdots \quad \ddots & \vdots \\ 0 & 0 \quad \ldots & H(k-1)\Phi(k-1,k) \\ 0 & 0 \quad \ldots & H(k) \end{bmatrix} \tag{5.59}$$

$$U_k \triangleq \begin{Bmatrix} U(1) \\ U(2) \\ \vdots \\ U(k-1) \end{Bmatrix} \tag{5.60}$$

Considering $V_k - M_k U_k$ as a "random disturbance" composed of both observation errors and "process noise"; and minimizing the criterion function:

$$(Y_k - A_k X(k))^t W_k (Y_k - A_k X(k)) \tag{5.61}$$

leads immediately to

$$\hat{X}_k(k) = P_k(k) A_k^t W_k \tilde{Y}_k \tag{5.62}$$

and

$$P_k(k) = (A_k^t W_k A_k)^{-1} \tag{5.63}$$

where, consistent with the results of chapter 2, the weight matrix is defined as

$$W_k \triangleq \left[E\left\{ (V_k - M_k U_k)(V_k - M_k U_k)^t \right\} \right]^{-1} \tag{5.64}$$

or

$$W_k^{-1} = \Lambda_{V_k V_k} - M_k Q(k) M_k^t \tag{5.65}$$

Now assume that new data

$$\tilde{Y}(k+1) = H(k+1)X(k+1) + V(k+1) \tag{5.66}$$

becomes available, it is desired to obtain a new estimate $\hat{X}_{k+1}(k+1)$ of $X(k+1)$, taking advantage of previous determinations of $\hat{X}_k(k)$ and $P_k(k)$. The previously derived homogeneous filter equations (5.42) apply with the exception of the covariance matrix update equation (5.42c) the results depend, however, upon knowledge of $P_k(k+1)$ whereas only $P_k(k)$ is available, a derivation of $P_k(k+1)$ similar to the homogeneous case follows.

From (5.48), it follows that

$$X(k) = \Phi(k,k+1)X(k+1) + \Phi(k,k+1)U(k), \tag{5.67}$$

which when substituted into the observation equations (5.58) yields

$$\tilde{Y}_k = A_k\Phi(k,k+1)X(k+1) - [\,M_k \vdots A_k\Phi(k,k+1)\,]\begin{bmatrix} U_k \\ \cdots \\ U(k)\end{bmatrix} + V_k \tag{5.68}$$

The estimate $\hat{X}_k(k+1)$ of $X(k+1)$ which minimizes the criterion function

$$\left(\tilde{Y}_k - A_k\Phi(k,k+1)X(k+1)\right)^t \Lambda_k^{-1}\left(\tilde{Y}_k - A_k\Phi(k,k+1)X(k+1)\right) \tag{5.69}$$

where

$$\Lambda_k = E\left\{[\,M_k \vdots A_k\Phi(k,k+1)\,]\begin{bmatrix} U_k \\ \cdots \\ U(k)\end{bmatrix}[U_k^t \vdots U^t(k)]\begin{bmatrix} M_k^t \\ \cdots\cdots \\ \Phi^t(k,k+1)A_k^t\end{bmatrix} + V_k V_k^t\right\} \tag{5.70}$$

or

$$\Lambda_k = M_k E\{U_k U_k^t\} M_k^t + E\{V_k V_k^t\} + A_k\Phi(,k+1)E\{U(k)U^t(k)\}\Phi^t(k,k+1)A_k^t \tag{5.71}$$

or, making use of (5.65) and (5.52)

$$\Lambda_k = W_k^{-1} + A_k\Phi(k,k+1)Q(k)\Phi^t(k,k+1)A_k^t \tag{5.72}$$

can be found to be

$$\hat{X}_k(k+1) = P_k(k+1)\Phi^t(k,k+1)A_k^t\Lambda_k^{-1}\tilde{Y}_k \tag{5.73}$$

where

$$P_k(k+1) = \left(\Phi^t(k,k+1)\, A_k^t \Lambda_k^{-1}\, A_k \Phi(k,k+1)\right)^{-1} \tag{5.74}$$

or, resubstituting (5.72)

$$P_k(k+1) = \left(\Phi^t(k,k+1)\, A_k^t\{W_k^{-1} + A_k\Phi(k,k+1)Q(k)\Phi^t(k,k+1)\, A_k^t\}^{-1} A_k\Phi(k,k+1)\right)^{-1} \tag{5.75}$$

which is an important intermediate form of the covariance update equation. An efficient scheme requiring inversion of a matrix of order m_k (the number of observations in the subset which becomes available at t_k) can be obtained by use of the matrix inversion identity of eqn. (1.84). Preparatory to these manipulations, it is convenient to compact (5.75) by the temporary definition

$$R \triangleq A_k\Phi(k,k+1) \tag{5.76}$$

so that (5.75) can be written as

$$P_k(k+1) = \left(R^t(W_k^{-1} + RQ(k)R^t)^{-1}R\right)^{-1} \tag{5.77}$$

or as

$$P_k(k+1) = \left(R^t\{W_k^{-1}(I + W_k RQ(k)R^t)\}^{-1}R\right)^{-1} \tag{5.78}$$

or as

$$P_k(k+1) = \left(R^t(I + W_k RQ(k)R^t)^{-1}W_k R\right)^{-1} \tag{5.79}$$

or as

$$P_k^{-1}(k+1) = R^t(I + W_k RQ(k)R^t)^{-1}W_k R \tag{5.80}$$

Equation (1.84) can be employed with

$$C = W_k R \tag{5.81}$$

and

$$D = Q(k)R^t \tag{5.82}$$

to obtain

$$P_k^{-1}(k+1) = R^t\{I - W_k R(I + Q(k) R^t W_k R)^{-1} Q(k) R^t\} W_k R \tag{5.83}$$

which can be manipulated as follows:

$$\begin{aligned} &P_k^{-1}(k+1) \\ &= R^t\{I - W_k R[((R^t W_k R)^{-1} + Q(k))(R^t W_k R)]^{-1} Q(k) R^t\} W_k R \\ &= R^t\{I - W_k R(R^t W_k R)^{-1}\{(R^t W_k R)^{-1} + Q(k)\}^{-1} Q(k) R^t\} W_k R \\ &= R^t W_k R - \{(R^t W_k R)^{-1} + Q(k)\}^{-1} Q(k) R^t W_k R \\ &= \{(R^t W_k R)^{-1} + Q(k)\}^{-1}\{\{(R^t W_k R)^{-1} + Q(k)\} R^t W_k R - Q(k) R^t W_k R\} \end{aligned}$$

Finally

$$P_k^{-1}(k+1) = \{(R^t W_k R)^{-1} + Q(k)\}^{-1}. \tag{5.84}$$

Resubstituting the definition (5.76), then (5.84) yields the final form of the covariance update equation as

$$P_k^{-1}(k+1) = [\{\Phi^t(k,k+1) A_k^t W_k A_k \Phi(k,k+1)\}^{-1} + Q(k)]^{-1}, \tag{5.85}$$

or

$$P_k(k+1) = \{\Phi^t(k,k+1) A_k^t W_k A_k \Phi(k,k+1)\}^{-1} + Q(k) \tag{5.86}$$

where $Q(k)$ is computed from the process noise $W(t)$ according to equations (5.54) and (5.55).

Equation (5.86) used in lieu of (5.42c), together with equations (5.42a), (5.42b) and (5.42d) constitute final form for a sequential Kalman filter for linear process-noise-forced dynamic systems. The problems of prediction and smoothing in the presence of process noise is treated by Gura(1966), and Jazwinski (1970).

The covariance update equation in the above classical form of the linear, sequential discrete Kalman filter is known to be particularily subject to roundoff and truncation errors which can (and often do) destroy the validity of the estimates obtained. A number of remedial actions and alternative developments have been stimulated by this numerical pitfall. This problem and possible cures are considered in chapter 6.

5.5 Optimal Continuous State Estimation

The generalization of the discrete measurement formulations developed in §5.4 to the corresponding continuous measurement formulations can be accomplished in a number of ways. The most rigorous approaches employ variational calculus to accomplish function space extremizations of least square, minimum variance, or maximum likelihood criterion functionals (Jazwinski (1970), for example). A less rigorous but more direct limiting process generalization of the discrete results is possible by simply allowing the time interval between measurements approach zero. Since this limiting process involves certain arguments that may not be satisfactory to mathematical purists, the interested reader is encouraged to consult the extensive literature on the subject.

The algebraic developments here follow the pattern of Gura (1966) and are most interesting in that they result directly from the least square or minimal variance concepts without requiring more involved function space arguments.

Assume a system under consideration is described by the homogeneous linear system of odes

$$\frac{dX(\tau)}{d\tau} = A(\tau)X(\tau) \tag{5.87}$$

and that observations are modeled by the linear algebraic equations

$$\tilde{Y}(\tau) = H(\tau)X(\tau) + V(\tau) \ . \tag{5.88}$$

Assume that the objective is to estimate the state $X(t)$ at $\tau = t$ based upon all measurements (5.88) made during the interval $t_0 \leq \tau \leq t$. From the results of §4.3 {equations (4.15) and (4.16)}, the solution of (5.87) can be written

$$X(\tau) = \Phi(\tau,t)X(t) \tag{5.89}$$

so that the observations (5.88) can be related to the state $X(t)$ as

$$\tilde{Y}(\tau) = H(\tau)\Phi(\tau,t)X(t) + V(\tau)\ . \tag{5.90}$$

Suppose that a large number of observations are made in subsets at discrete instants of time $\tau_1, \tau_2, \cdots, \tau_k$, further suppose for simplicity that the time interval $\Delta\tau$ between measurements is constant. Then the k subsets of observation equations are

$$\begin{aligned} \tilde{Y}(\tau_1) &= H(\tau_1)\Phi(\tau_1,\tau_k)X(\tau_k) + V(\tau_1) \\ \tilde{Y}(\tau_2) &= H(\tau_2)\Phi(\tau_2,\tau_k)X(\tau_k) + V(\tau_2) \\ &\vdots \\ \tilde{Y}(\tau_k) &= H(\tau_k)\Phi(\tau_k,\tau_k)X(\tau_k) + V(\tau_k) \end{aligned} \tag{5.91}$$

Equations (5.91) are of the same structure as the observation equations considered in §1.2 {eqns. (1.17)}, for which the optimal estimate (using either least square or minimal variance criterions) can be written by analogy with equation (1.30) as

$$\hat{X}(\tau_k) = \left(\sum_{i=1}^{k} \Phi^t(\tau_i,\tau_k)H^t(\tau_i)W(\tau_i)H(\tau_i)\Phi(\tau_i,\tau_k) \right)^{-1} \sum_{i=1}^{k} \Phi^t(\tau_i,\tau_k)H^t(\tau_i)W(\tau_i)\tilde{Y}(\tau_k) \tag{5.92}$$

The ith weighting matrix must be taken as inverses of the ith measurement error covariance matrix, for a minimum variance estimate of X (see chapter 2).

We now assume that the number of observation subsets becomes arbitrarily large, while the intervening time interval $\Delta\tau$ becomes arbitrarily small. Then, in the limit, equation (8.6) becomes (letting $\tau_k \equiv t$)

$$\hat{X}(t) = \lim_{\substack{k\to\infty \\ \Delta\tau\to 0}} \left\{ \left(\sum_{i=1}^{k} \Phi^t(\tau_i,t_k)H^t(\tau_i)W(\tau_i)H(\tau_i)\Phi(\tau_i,t_k) \right)^{-1} \sum_{i=1}^{k} \Phi(\tau_i,t_k)H^t(\tau_i)W(\tau_i)\tilde{Y}(t_k) \right\}$$

or

$$\hat{X}(t) = \lim_{\substack{k\to\infty \\ \Delta\tau\to 0}} \left\{ \left(\sum_{i=1}^{k} \Phi^t(\tau_i,t_k)H^t(\tau_i)W(\tau_i)H(\tau_i)\Phi(\tau_i,t_k)\Delta\tau \right)^{-1} \sum_{i=1}^{k} \Phi\,(\tau_i,t_k)H^t(\tau_i)W(\tau_i)\tilde{Y}(\tau_i)\Delta\tau \right\}$$

or

$$\hat{X}(t) = \left(\int_{t_o}^{t} \Phi^t(\tau,t)H^t(\tau)W(\tau)H(\tau)\Phi(\tau,t)d\tau \right)^{-1} x \int_{t_o}^{t} \Phi^t(\tau,t)H^t(\tau)W(\tau)\tilde{Y}(\tau)d\tau \tag{5.93}$$

Compare this result with

$$\hat{X}(t) = \{\Phi^t(t,t)H^t(t)W(t)H(t)\Phi(t,t)\}^{-1}\Phi^t(t,t)H^t(t)W(t)\tilde{Y}(t) \tag{5.94}$$

which immediately results from minimizing "errors after the solution" of (5.88) at any given instant. Equations (5.93) is the estimate of $\hat{X}(t)$ based upon *all* measurements up to time t, while equation (5.94) is the estimate of $\hat{X}(t)$ based *only* upon the subset of observations instantaneously available at time t.

The general result (5.93) can be derived directly using calculus of variations to determine the estimate $\hat{X}(t)$ which minimizes the integral weighted sum of squares

$$\phi = \int_{t_o}^{t} \{\tilde{Y}(\tau) - H(\tau)\Phi(\tau,t)X(\tau)\}^t W(\tau)\{\tilde{Y}(\tau) - H(\tau)\Phi(\tau,t)X(\tau)\}d\tau \tag{5.95}$$

This criterion is clearly the function space analog of the parameter space least squares criterion.

Since equation (5.93) requires inversion (at each integration step) an n x n matrix, it is not very efficient for large n. After considerable manipulations, we will find it possible to rearrange the solution so that an inversion of an m x m matrix is required, where m is the number of observations in the subset of measurements available at a given instant of time.

Analogous to the discrete covariance matrix, we define

$$P^{-1}(t) \triangleq \int_{t_o}^{t} \Phi^t(\tau,t)H^t(\tau)W(\tau)H(\tau)\Phi(\tau,t)d\tau \tag{5.96}$$

so that the solution (5.93) becomes

$$\hat{X}(t) = P(t) \int_{t_o}^{t} \Phi^t(\tau,t)H^t(\tau)W(\tau)\tilde{Y}(\tau)d\tau \tag{5.97}$$

Differentiating (5.96) and (5.97), we obtain

$$\begin{aligned}\frac{dP(t)}{dt} = \Bigg\{ &\int_{t_o}^{t} \frac{d\Phi^t}{dt}(\tau,t)H^t(\tau)W(\tau)H(\tau)\Phi(\tau,t)d\tau \\ &+ \int_{t_o}^{t} \Phi^t(\tau,t)H^t(\tau)W(\tau)H(\tau)\frac{d\Phi(\tau,t)}{dt}d\tau \\ &+ \Phi^t(t,t)H^t(t)W(t)H(t)\Phi(t,t)\Bigg\}^{-1}\end{aligned} \tag{5.98}$$

and

$$\begin{aligned}\frac{d\hat{X}(t)}{dt} &= \frac{dP(t)}{dt}\int_{t_o}^{t} \Phi^t(\tau,t)H^t(\tau)W(\tau)\tilde{Y}(\tau)d\tau \\ &+P(t)\int_{t_o}^{t} \frac{d\Phi^t(\tau,t)}{dt}H^t(\tau)W(\tau)\tilde{Y}(\tau)d\tau + P(t)\Phi^t(t,t)H^t(t)W(t)\tilde{Y}(t)\ .\end{aligned} \tag{5.99}$$

For any nonsingular matrix M(t), differentiation of $M(t)M^{-1}(t) = I$ immediately yields the identity

$$\frac{dM^{-1}(t)}{dt} = M^{-1}(t)\,\frac{dM(t)}{dt}\,M^{-1}(t) \tag{5.100}$$

which, for $M(t) = \Phi(t,\tau)$ yields

$$\frac{d\Phi^{-1}(t,\tau)}{dt} = -\Phi^{-1}(t,\tau)\,\frac{d\Phi(t,\tau)}{dt}\,\Phi^{-1}(t,\tau)\ . \tag{5.101}$$

Making use of the property $\Phi^{-1}(t,\tau) = \Phi(\tau,t)$, then (5.101) becomes

$$\frac{d\Phi(\tau,t)}{dt} = -\Phi(\tau,t)\,\frac{d\Phi(t,\tau)}{dt}\,\Phi(\tau,t)\ , \tag{5.102}$$

and since

$$\frac{d\Phi(t,\tau)}{dt} = A(t)\Phi(t,\tau)\ ,$$

then (5.102) becomes the *"adjoint equation"*

$$\frac{d\Phi(\tau,t)}{dt} = -\Phi(\tau,t)A(t). \tag{5.103}$$

Using this result in (5.98) yields

$$\begin{aligned}\frac{dP(t)}{dt} = \Big\{ &-A^t(t)\int_{t_o}^{t} \Phi^t(\tau,t)H^t(\tau)W(\tau)H(\tau)\Phi(\tau,t)d\tau \\ &- \int_{t_o}^{t} \Phi^t(\tau,t)H^t(\tau)W(\tau)H(\tau)\Phi(\tau,t)d\tau\ A(t) \\ &+H^t(t)W(t)H(t)\Big\}^{-1}\end{aligned} \tag{5.104}$$

Using Equation (5.96), then equation (5.104) can be written as

$$\frac{dP^{-1}(t)}{dt} = -A^t(t)P^{-1}(t)-P^{-1}(t)A(t)+H^t(t)W(t)H(t) \tag{5.105}$$

Substitution of (5.97) and (5.103) into (5.99) yields

$$\frac{d\hat{X}(t)}{dt} = \frac{dP(t)}{dt}\,P^{-1}(t)\hat{X}(t)-P(t)A^t(t)P^{-1}(t)\hat{X}(t)+P(t)H^t(t)W(t)\tilde{Y}(t) \tag{5.106}$$

Notice that equation (5.100) with $M = P^{-1}(t)$ yields

$$\frac{dP(t)}{dt} = -P(t)\,\frac{dP^{-1}(t)}{dt}\,P(t) \tag{5.107}$$

which, upon substituting from (5.105) yields the *matrix Ricatti equation*

$$\frac{dP(t)}{dt} = P(t)A^t(t) + A(t)P(t) - P(t)H^t(t)W(t)H(t)P(t) \quad . \tag{5.108}$$

Equations (5.106) now become

$$\frac{d\hat{X}(t)}{dt} = A(t)\hat{X}(t) + P(t)H^t(t)W(t)\{\tilde{Y}(t) - H(t)\hat{X}(t)\} \tag{5.109}$$

Equations (5.108) and (5.109) are the standard form {Kalman (1960)} of the *unforced Kalman Filter* for continuous estimation of linear systems.

The initial conditions on (5.108) and (5.109) are a source of some theoretical but little practical difficulty. From (5.93), at $t = t_0$, it is apparent that $P(t_0) = \infty$ and $\hat{X}(t_0)$ is indeterminate. This is consistent with the manner in which the derivation was carried out, i.e., it was implicitely assumed that *zero* information on X and P were available prior to t_0, therefore these initial conditions are consistent. In practice, the initial conditions could be obtained either from inversion of observations at discrete times {e.g., from equation (5.95)} or apriori estimates could be employed.

5.6 Sequential Estimation of Nonlinear Dynamical Systems

We consider now the class of systems whose state history is governed by a system of nonlinear ordinary differential equations

$$\dot{X} = F(t,X,U) \tag{5.110}$$

and whose measurements are modeled by a system of nonlinear algebraic equations

$$\tilde{Y}(t) = G(X(t)) + V \quad . \tag{5.111}$$

One approach to applying the linear estimation methods of §5.5 and 6.6 to the present system is to introduce a typical *nominal* trajectory $X_n(t)$ and linearize equations (5.110) and (5.111) in the vicinity of the nominal state history as

$$\frac{d}{dt}(\delta X) = A\ \delta X + B\ \delta U \tag{5.112}$$

and

$$\delta Y \cong H\ \delta X + V, \tag{5.113}$$

where, in this case, the δ's denote instantaneous "departures from nominal" as

$$\delta X = X(t) - X_n(t) \tag{5.114}$$

$$\delta Y = \tilde{Y}(t) - Y_n(t) \tag{5.115}$$

and

$$A = \left.\frac{\partial F}{\partial X}\right|_n \tag{5.116}$$

$$B = \left.\frac{\partial F}{\partial U}\right|_n \tag{5.117}$$

$$H = \left.\frac{\partial G}{\partial X}\right|_n \quad . \tag{5.118}$$

Then, for example, the discrete time linear Kalman Filter equations (5.42) and (5.86) can be directly applied to sequential estimation of $\delta X(t_k)$ as

$$\delta\hat{X}_{k+1}(k+1) = \Phi(k+1,k)\ \delta\hat{X}_k(k) + K(k+1)\{\delta\tilde{Y}(k+1) - H(k+1)\Phi(k+1,k)\delta X_k(k)\} \tag{5.119a}$$

and

$$P_{k+1}(k+1) = \{I - K(k+1)H(k+1)\}\ P_k(k+1) \tag{5.119b}$$

where

$$P_k(k+1) = \Phi(k+1,k)P_k(k)\Phi^t(k+1,k) + Q(k) \tag{5.119c}$$

and

$$K(k+1) = P_k(k+1)H^t(k+1)\{\Lambda_{v_{k+1}v_{k+1}} + H(k+1)P_k(k+1)H^t(k+1)\}^{-1}. \tag{5.119d}$$

This approach applies only for linearizable small displacements $\delta X(t)$ from the nominal trajectory; this restriction proves especially damaging for highly nonlinear applications. This drawback of (5.119) motivated the development {Schutz (1973), and Jaswinski (1970)} of the so-called *extended Kalman Filter* which eliminates the major source of errors due to neglected nonlinearities Observe that the first term $\Phi(t_{k+1},t_k)\ \delta\hat{X}_k(t_k)$ of equation (5.119) is the forward integration of the linearized departure motion equations (5.112), based upon the optimal estimate of the departure at time t_k. Equations (5.119) can be greatly improved ("extended") by simply replacing all linear approximations to departure motion by full nonlinear integrations of equations (5.110) between each measurement time. Adopting the notation

$\bar{X}_k(t_{k+1})$ = nonlinear integrated estimate of the state at time t_{k+1} based upon integrating equations (5.110) with the optimal state estimate $\hat{X}_k(t_k)$,

then, as is justified in detail by Jazwinski (1970), the "extended" Kalman Filter state update equation is

$$\hat{X}_{k+1}(k+1) = \bar{X}_k(k+1) + K(k+1)\ \{\tilde{Y}(k+1) - G(\bar{X}_k(k+1))\}, \tag{5.120}$$

equations (5.119b-d) still hold.

Observe that the covariance propagation implicit in the extended formulation is still based upon the linear departure motion approximation, but the nominal trajectory is revised after each state update (5.120) and thus the linearizations have a greater liklihood of being valid. Schutz (1973) considers interesting extensions and applications of the extended Kalman formulation to estimate the orbit of a Lunar satellite in the presence of gravitational model process noise. Example applications of the formulations (5.119) and (5.120) are presented in chapter 6.

5.7 *Remarks*

The developments of the present chapter depart from tradition in several respects. Most significantly, limiting arguments are used to obtain the continuous estimation results from the discrete case, rather than employing function space and calculus of variations to rigorously derive the results. This approach, after Gura (1967), accomplishes two important objectives; (1) it establishes clearly the connection of the discrete and continuous formulations, and (2) it makes the discussion fully accessible by most senior undergraduate engineering students. The loss of rigor and esthetic value of the developments is perhaps significant, but the student who continues to study these topics and co-requisite studies in stochastic processes, variational calculus, and control theory, will have excellent preparation based upon the developments herein.

5.8 *Exercises*

5.1 Given the coupled weakly non-linear oscillators

$$\ddot{x} = -\omega_1^2 x + \varepsilon xz + A\cos\Omega_1 t$$

$$\ddot{z} = -\omega_2^2 z - \varepsilon xz + B\cos\Omega_2 t$$

and the measurement model equation

$$\tilde{y}(t) = Cx + Dz + v,$$

where $\omega_1^2, \omega_2^2, A, B, C, D$, and ε are scalar constants, and $E(v) = 0$, $E(v^2(t_j)) = q$, $E(v(t_i)v(t_j)) = 0$.

Consider the following estimation problems:

(A) The model parameters $\{\omega_1^2, \omega_2^2, \Omega_1^2, \Omega_2^2, A, B, C, D, \varepsilon\}$ are given constants, y can be measured at m discrete instants $(t_1, t_2, \cdots, t_m)$; it is desired to estimate the initial state vector

$$X(t_o) = \begin{Bmatrix} x(t_o) \\ y(t_o) \\ \dot{x}(t_o) \\ \dot{y}(t_o) \end{Bmatrix} \quad \text{, given an initial estimate}$$

$\hat{X}_a(t_o)$, and associated covariance matrix $P(t_o)$.

(B) The nine model parameters are uncertain, y can be measured at m discrete instants, it is desired to estimate the initial state vector $\{x(t_o), y(t_o), \dot{x}(t_o), \dot{y}(t_o)\}$ and the nine model parameters $\{\omega_1^2, \omega_2^2, \Omega_1^2, \Omega_2^2, A, B, C, D, \varepsilon\}$, given apriori estimates and an associated covariance matrices.

(C) The nine model parameters are assumed perfectly known, it is desired to estimate the instantaneouse state

$$X^t(j+1) = \{x(t_{j+1}) y(t_{j+1}) \dot{x}(t_{j+1}) \dot{y}(t_{j+1})\}$$

immediately upon receipt of each new measurement $\tilde{y}(t_{j+1})$.

Using the methods of the present chapter, formulate minimal variance estimation algorithms for the above three problems. If time permits, implement the three algorithms as computer programs and study the performance of the algorithms (use synthetic measured data, generated by adding zero mean normally distributed random numbers to perfect calculated y-values, see how well the true state and model parameter values are recovered).

5.9 References for Chapter 5

5.1 Schutz, B.E., "Analytical Approach to Orbit Determination in the Presence of Model Errors", Amer. Inst. of Aeronautics and Astronautics paper #73-170, presented to the 11th AIAA Aerospace Sciences Meeting, Wash., D.C. (1973).

5.2 Jazwinski, A.H., Stochastic Processes and Filtering Theory, Academic Press, N.Y., ch 8 and 9, (1970).

5.3 Gura, I.A., "An Algebraic Approach to Optimal Estimation", Hughes Space Systems Report No. SSD 70072R, ElSegundo, Ca. (1967).

5.4 Kalman, R.E., "A New Approach to Linear Filtering and Prediction Problems", Trans. ASME, Vol 82D, p35 (1960).

6

Estimation of Dynamical Systems: Applications

6.1 Projectile Trajectory Estimation

We begin this chapter by considering an idealization of the radar tracking/trajectory estimation problem. Referring to Fig. 6.1, the radar (located at the origin) is assumed capable of measuring the slant range (ρ), azimuth (ϕ), and elevation (θ).

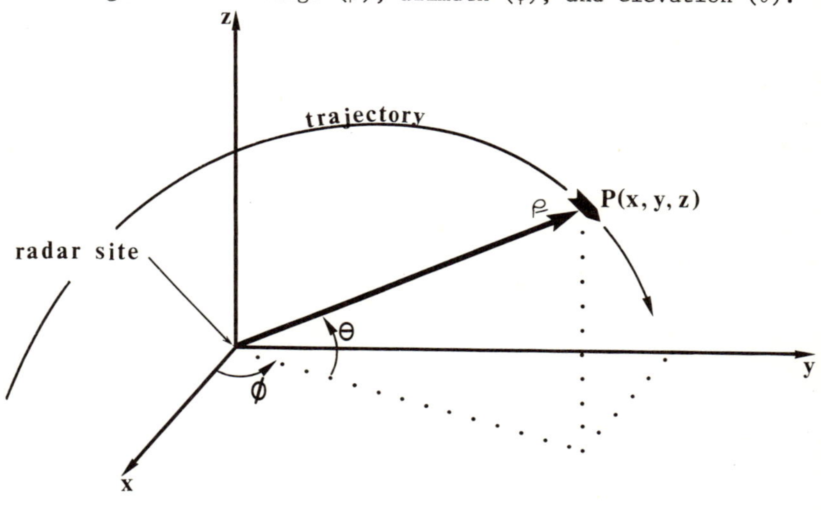

Figure 6.1 Radar Measurement Geometry

It is clear from the geometry that perfect data satisfy the nonlinear algebraic equations

$$\rho(t) = \sqrt{x^2(t) + y^2(t) + z^2(t)} \tag{6.1}$$

$$\phi(t) = \tan^{-1}\left(\frac{y(t)}{x(t)}\right) \quad , \; -\pi < \phi < \pi \tag{6.2}$$

and

$$\theta(t) = \sin^{-1}\left(\frac{z(t)}{\rho(t)}\right) \quad , \; -\frac{\pi}{2} < \theta < \frac{\pi}{2}. \tag{6.3}$$

For simplicity, assume that the only force acting on the projectile is a constant gravity force (mg) so that the state dynamics are governed by the three differential equations

$$\begin{aligned} \ddot{x} &= 0 \\ \ddot{y} &= 0 \\ \ddot{z} &= -g \end{aligned} \tag{6.4}$$

The integration of equations (6.4) is near-trivial, and as established previously (example 4.4), the solution for the state vector can be written as

$$\begin{Bmatrix} x_1(t) \\ x_2(t) \\ x_3(t) \\ x_4(t) \\ x_5(t) \\ x_6(t) \end{Bmatrix} = \{\Phi(t,t_j)\} \begin{Bmatrix} x_1(t_j) \\ x_2(t_j) \\ x_3(t_j) \\ x_4(t_j) \\ x_5(t_j) \\ x_6(t_j) \end{Bmatrix} + \begin{Bmatrix} 0 \\ 0 \\ -\frac{1}{2}g(t-t_j)^2 \\ 0 \\ 0 \\ -g(t-t_j) \end{Bmatrix} \tag{6.5}$$

where the state vector components are

$$x_1 \equiv x, \; x_2 \equiv y, \; x_3 \equiv z, \; x_4 \equiv \dot{x}, \; x_5 \equiv \dot{y}, \; x_6 \equiv \dot{z}$$

and the state transition matrix is

$$\Phi(t,t_j) \equiv \begin{bmatrix} 1 & 0 & 0 & \Delta t & 0 & 0 \\ 0 & 1 & 0 & 0 & \Delta t & 0 \\ 0 & 0 & 1 & 0 & 0 & \Delta t \\ 0 & 0 & 0 & 1 & 0 & 0 \\ 0 & 0 & 0 & 0 & 1 & 0 \\ 0 & 0 & 0 & 0 & 0 & 1 \end{bmatrix} \tag{6.6}$$

with $\Delta t = t-t_j$, for t_j an arbitrary reference time.

We now consider two estimation problems:

Problem 6.1A Initial State Estimation

Problem 6.1B Sequential State Estimation

We will define each problem completely and carry out specific numerical solutions.

Measured data (Table 6.1) were simulated by assuming the *true* initial state

$$\begin{aligned} x(0) &= -10000 \text{ m} \\ y(0) &= -\ 3000 \text{ m} \\ z(0) &= 100 \text{ m} \\ \dot{x}(0) &= 500 \text{ m/sec} \\ \dot{y}(0) &= 100 \text{ m/sec} \\ \dot{z}(0) &= 800 \text{ m/sec} \end{aligned}$$

and calculating perfect values for (ρ,ϕ,θ) from equations (6.1), (6.2), and (6.3). The simulated measurements $(\tilde{\rho},\tilde{\phi},\tilde{\theta})$ of Table 6.1 were determined from the corresponding perfect data (ρ,ϕ,θ) by

$$\tilde{\rho}(t_j) = \rho(t_j) + V_\rho(t_j) \tag{6.7a}$$

$$\tilde{\phi}(t_j) = \phi(t_j) + V_\phi(t_j) \tag{6.7b}$$

$$\tilde{\theta}(t_j) = \theta(t_j) + V_\theta(t_j) \tag{6.7c}$$

where the measurement errors V_ρ, V_ϕ, V_θ were taken as Gaussian random numbers. Adopting the notation

$$V^t = \{V_\rho \ V_\phi \ V_\theta\}$$

then the first two moments of the simulated measurement errors were

$$E\{V\} = 0 \qquad \text{for all } t$$

$$E\{V(t_i)V^t(t_j)\} = 0 \ , \ i \neq j$$

and

TABLE 6.1

SIMULATED RADAR MEASUREMENTS OF PROJECTILE MOTION

Time(sec)	$\tilde{\rho}$(m)	$\tilde{\phi}$(radians)	$\tilde{\theta}$(radians)
0	.104574E+05	-.285954E+01	.545636E-02
2	.962752E+04	-.284103E+01	.184077E+00
4	.896865E+04	-.284196E+01	.361298E+00
6	.862802E+04	-.280326E+01	.530708E+00
8	.839271E+04	-.280664E+01	.706333E+00
10	.845331E+04	-.275610E+01	.876141E+00
12	.867152E+04	-.270632E+01	.103973E+01
14	.888497E+04	-.264903E+01	.117680E+01
16	.923345E+04	-.253415E+01	.131399E+01
18	.939355E+04	-.227303E+01	.140669E+01
20	.976026E+04	-.157178E+01	.148383E+01
22	.100144E+05	-.673696E+00	.144218E+01
24	.101641E+05	-.290519E+00	.139101E+01
26	.104982E+05	-.127255E+00	.125905E+01
28	.107396E+05	-.459796E-01	.119445E+01
30	.108845E+05	-.110464E-01	.110751E+01
32	.110675E+05	.438265E-01	.983099E+00
34	.112371E+05	.653596E-01	.896168E+00
36	.113817E+05	.595262E-01	.781255E+00
38	.115405E+05	.866973E-01	.674190E+00

TABLE 6.2

LEAST SQUARE DIFFERENTIAL CORRECTION SOLUTION 6.1A

Iteration	$x(t_0)$	$y(t_0)$	$z(t_0)$	$\dot{x}(t_0)$	$\dot{y}(t_0)$	$\dot{z}(t_0)$	Weighted Sum Square of Residuals
0	-5000.	-1500.	50.	250.0	50.0	400.0	5.6×10^6
1	-14152.	-3531.	3821.	727.6	109.2	348.7	2.0×10^6
2	-13879.	-2295.	-997.	760.2	36.5	814.0	5.8×10^5
3	-10263.	-3562.	-306.	502.9	119.1	839.6	2.9×10^4
4	-10048.	-2973.	179.	503.0	98.5	797.9	6.0×10^2
5	-10019.	-2986.	164.	500.9	99.6	798.5	5.3×10^2
6	-10020.	-2986.	164.	500.9	99.6	798.5	5.3×10^2

Converged State Covariance Matrix:

$$(A^tWA)^{-1} = \begin{bmatrix} 584 & -16 & 56 & -28 & 2 & 2 \\ -16 & 756 & 89 & 2 & -33 & -3 \\ 56 & 89 & 961 & -1 & -4 & -38 \\ -28 & 2 & -1 & 1.5 & -0.2 & -0.2 \\ 2 & -33 & -4 & -0.2 & 1.6 & 0.2 \\ 2 & -3 & -38 & -0.2 & 0.2 & 1.9 \end{bmatrix}$$

$$E\{V(t_j)V^t(t_j)\} = \begin{bmatrix} \sigma_\rho^2 & 0 & 0 \\ 0 & \sigma_\phi^2 & 0 \\ 0 & 0 & \sigma_\theta^2 \end{bmatrix} = \begin{bmatrix} 2500 & 0 & 0 \\ 0 & 0.0001 & 0 \\ 0 & 0 & 0.0001 \end{bmatrix}. \tag{6.8}$$

Thus each of the variables ρ, ϕ, θ are corrupted by an independent sequence of *white noise* to simulate measurement errors.

Solution 6.1A Initial Condition Estimation via Batch Least Square Differential Correction (LSDC)

The conventional least square differential correction algorithm

$$\hat{X}^{(k+1)} = \hat{X}^{(k)} + \{(A^t WA)^{-1} A^t W\Delta Y\}^{(k)} \tag{6.9}$$

is employed to estimate the vector of initial projectile conditions

$$X^t = \{x(0)\ y(0)\ \cdots\ \dot{z}(0)\} \tag{6.10}$$

where $t_1 = 0$ is an arbitrary reference time, and where

$$\underset{60\times 1}{\tilde{Y}} = \begin{bmatrix} \rho(t_1) \\ \phi(t_1) \\ \theta(t_1) \\ \cdot \\ \cdot \\ \cdot \\ \rho(t_{20}) \\ \phi(t_{20}) \\ \theta(t_{20}) \end{bmatrix}, \quad \underset{60\times 6}{A} = \left(\frac{\partial y_j}{\partial x_i} \bigg|_C \right) \tag{6.11, 6.12}$$

$$\underset{60\times 60}{W} = \begin{bmatrix} \frac{1}{\sigma_{\rho_1}^2} & & & & & & 0 \\ & \frac{1}{\sigma_{\phi_1}^2} & & & & & \\ & & \frac{1}{\sigma_{\theta_1}^2} & & & & \\ & & & \ddots & & & \\ & & & & \frac{1}{\sigma_{\rho_{20}}^2} & & \\ & & & & & \frac{1}{\sigma_{\phi_{20}}^2} & \\ 0 & & & & & & \frac{1}{\sigma_{\theta_{20}}^2} \end{bmatrix} \tag{6.14}$$

Adopting the arbitrary (off by 50%) starting estimates of the unknown initial state vector:

$$X^{(0)} = \begin{Bmatrix} -5000 \\ -1500 \\ 50 \\ 250 \\ 50 \\ 400 \end{Bmatrix} \tag{6.15}$$

then the LSDC algorithm (6.9) results in the convergence history summarized in Table 6.2.

Solution 6.1B Sequential State Estimation via an Extended Kalman Filter Algorithm

The state vector notation adopted here is

$$\hat{X}_j(k) = \begin{Bmatrix} x_j(t_k) \\ y_j(t_k) \\ z_j(t_k) \\ \dot{x}_j(t_k) \\ \dot{y}_j(t_k) \\ \dot{z}_j(t_k) \end{Bmatrix} = \text{optimal estimates of} \begin{Bmatrix} x \\ y \\ z \\ \dot{x} \\ \dot{y} \\ \dot{z} \end{Bmatrix} \text{at time } t_k \tag{6.16}$$

based upon the first j data subsets.

The extended Kalman filter algorithm eqns.{(5.120) and (5.119A)-(5.119d)} adopted for this solution is

$$\hat{X}_{k+1}(k+1) = \bar{X}_k(k+1) + K_k(k+1)\{\tilde{Y}(k+1) - \bar{Y}_k(k+1)\} \tag{6.17a}$$

$$P_{k+1}(k+1) = \{I - K(k+1)H(k+1)\}P_k(k+1) \tag{6.17b}$$

$$P_k(k+1) = \Phi(k+1,k)\ P_k(k)\ \Phi^t(k+1,k) \tag{6.17c}$$

$$K(k+1) = P_k(k+1)\ H^t(k+1)\ \{\Lambda_{v_{k+1}v_{k+1}} + H(k+1)\ P_k(k+1)\ H^t(k+1)\}^{-1} \tag{6.17d}$$

where

$$\bar{X}_k(k+1) = \Phi(k+1,k)\hat{X}_k(k) + \begin{Bmatrix} 0 \\ 0 \\ -g(t_{k+1} - t_k)^{2}/2 \\ 0 \\ 0 \\ -g(t_{k+1} - t_k) \end{Bmatrix} \tag{6.18}$$

$$\bar{Y}_k(k+1) = \begin{Bmatrix} \rho_k(k+1) \\ \phi_k(k+1) \\ \theta_k(k+1) \end{Bmatrix} \tag{6.19}$$

$$\rho_k(k+1) = \{x_k^2(k+1) + y_k^2(k+1) + z_k^2(k+1)\}^{\frac{1}{2}} \tag{6.20}$$

$$\phi_k(k+1) = \tan^{-1}\left(\frac{y_k(k+1)}{x_k(k+1)}\right) \tag{6.21}$$

$$\theta_k(k+1) = \sin^{-1}\left(\frac{z_k(k+1)}{\rho_k(k+1)}\right) \tag{6.22}$$

$$\tilde{Y}(k+1) = \text{measured values of } \begin{Bmatrix} \rho \\ \phi \\ \theta \end{Bmatrix} \text{ at time } t_{k+1} \text{ (Table 6.1).} \tag{6.23}$$

$$H(k+1) = \begin{bmatrix} \left.\frac{\partial\rho}{\partial x}\right|_{\bar{X}_k(k+1)} & \left.\frac{\partial\rho}{\partial y}\right|_{\bar{X}_k(k+1)} & \left.\frac{\partial\rho}{\partial z}\right|_{\bar{X}_k(k+1)} & 0 & 0 & 0 \\ \left.\frac{\partial\phi}{\partial x}\right|_{\bar{X}_k(k+1)} & \left.\frac{\partial\phi}{\partial y}\right|_{\bar{X}_k(k+1)} & \left.\frac{\partial\phi}{\partial z}\right|_{\bar{X}_k(k+1)} & 0 & 0 & 0 \\ \left.\frac{\partial\theta}{\partial x}\right|_{\bar{X}_k(k+1)} & \left.\frac{\partial\theta}{\partial y}\right|_{\bar{X}_k(k+1)} & \left.\frac{\partial\theta}{\partial z}\right|_{\bar{X}_k(k+1)} & 0 & 0 & 0 \end{bmatrix} \tag{6.24}$$

$$\Phi(k+1,k) = \begin{bmatrix} 1 & 0 & 0 & (t_{k+1} - t_k) & 0 & 0 \\ 0 & 1 & 0 & 0 & (t_{k+1} - t_k) & 0 \\ 0 & 0 & 1 & 0 & 0 & (t_{k+1} - t_k) \\ 0 & 0 & 0 & 1 & 0 & 0 \\ 0 & 0 & 0 & 0 & 1 & 0 \\ 0 & 0 & 0 & 0 & 0 & 1 \end{bmatrix} \tag{6.25}$$

and

$$\Lambda_{v_{k+1}v_{k+1}} = \begin{bmatrix} \sigma^2_{\rho_{k+1}} & & 0 \\ & \sigma^2_{\phi_{k+1}} & \\ 0 & & \sigma^2_{\theta_{k+1}} \end{bmatrix} \tag{6.26}$$

to start the estimation process, the state estimates (as for LSDC of solution 6.1A, off by 50%) were set to equation (6.15), and the initial state covariance matrix was taken as

$$P_0(0) = \begin{bmatrix} 10^6 & 0 & 0 & 0 & 0 & 0 \\ 0 & 10^6 & 0 & 0 & 0 & 0 \\ 0 & 0 & 10^6 & 0 & 0 & 0 \\ 0 & 0 & 0 & 10^6 & 0 & 0 \\ 0 & 0 & 0 & 0 & 10^6 & 0 \\ 0 & 0 & 0 & 0 & 0 & 10^6 \end{bmatrix} \tag{6.27}$$

The twenty sets of ρ, ϕ, θ observations (Table 6.1) were processed by the above algorithm. The program was structured to operate in two modes:

Mode 1 A non-iterative forward sweep in which the twenty data sets were processed once; this results in state and covariance estimates at $t = 0, 2, 4, \cdots 38$ seconds.

Mode 2 An iteration procedure in which the data is processed forward to $t = 38$, then backward to $t = 0$. Each backward pass is initiated with the $t = 38$ state estimates (of the preceding forward pass) and with equation (6.27) as the covariance matrix. Each forward pass (after the first) is initiated with the preceding $t = 0$ state estimate and (6.27) as the covariance matrix. The process terminates when state convergence at $t = 0$ occurs (usually one or two iterations).

TABLE 6.3

ACTUAL POSITION ESTIMATION ERRORS FOR KALMAN FILTER (MODE 1) SOLUTION 6.1B

Time	ΔX	σ_x	ΔY	σ_y	ΔZ	σ_z	$\lvert\Delta\rho\rvert$	σ_ρ
0-	5000	1000	1500.	1000.	-50.	1000.	5220.	1000.
0+	-18.	50.	46.	52.	21.	52.	54.	50.
2	-85.	56.	-20.	96.	165.	99.	187.	62.
4	-45.	54.	95.	77.	141.	80.	176.	60.
6	-24.	51.	-16.	62.	93.	66.	98.	57.
8	-30.	50.	54.	52.	26.	56.	66.	53.
10	-14.	50.	3.	43.	14.	49.	20.	48.
12	10.	46.	-32.	37.	54.	43.	64.	44.
14	6.	44.	-26.	32.	58.	39.	64.	39.
16	20.	41.	3.	31.	97.	35.	99.	35.
18	-6.	31.	-1.	33.	54.	32.	54.	32.
20	0.	10.	7.	23.	55.	29.	55.	29.
22	3.	10.	1.	14.	45.	28.	45.	28.
24	1.	12.	3.	14.	12.	27.	12.	27.
26	7.	14.	4.	14.	9.	26.	12.	26.
28	10.	17.	5.	15.	23.	26.	26.	25.
30	11.	19.	0.	17.	30.	26.	32.	24.
32	18.	21.	4.	18.	31.	26.	36.	24.
34	20.	22.	8.	20.	37.	26.	43.	25.
36	27.	24.	0.	21.	28.	26.	39.	25.
38	19.	25.	-2.	23.	14.	27.	24.	26.

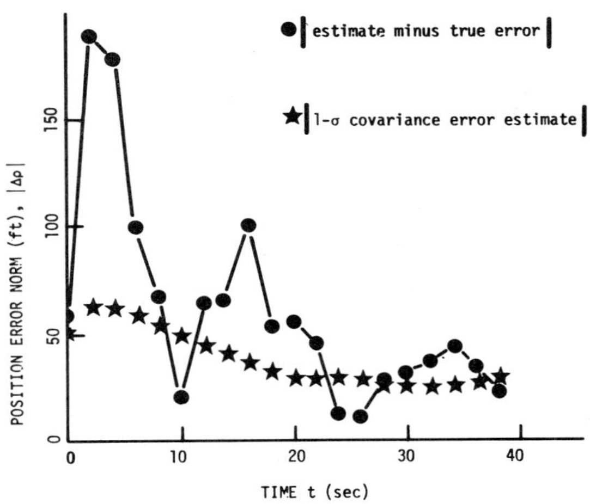

Figure 6.2 Actual Position Errors versus 1-σ Covariance Estimates for the Kalman Filter Solution 6.1B(Mode 1)

The Mode 1 results are summarized in Table 6.3 and Figure 6.2. Observe that, after an initial transient (due to the poor starting estimates), that 3σ covariance estimates are usually conservative estimates of the actual estimation errors. This indicates that linear error theory is reasonably valid for this application. The fact that the projectile makes its closest approach around ten seconds causes an anomalous minimum in the estimation errors (since the angle observation errors cause the least position uncertainty there). Thereafter, it is interesting to observe that the competing effects of more observations and increasing range causes a mild "roller coaster" oscillation in error magnitudes and associated covariance estimates. Single and double precision calculations were compared to verify that the single precision calculations of this example do not suffer from numerical instabilities of the type discussed in §6.2.

The Mode 2 (iterative extended Kalman filter) results are summarized in Table 6.4 in which the state and covariance estimates are given at the following times:

- $t = 0-$ (starting estimates)
- $t = 0+$ (first observation time)
- $t = 38$ (end of first forward pass)
- $t = 0$ (end of first backward pass)
- $t = 0$ (end of second backward pass)

Comparison of the converged ($t = 0$) estimates of Table 6.4 with the batch least square differential correction results (Table 6.2), one concludes that the "batch" and "iterative sequential" estimation yielded (to well within the observation noise level) identical converged solutions. It is significant to note that while the state estimate disagreement is negligible, there are slight disagreements between the associated covariance matrices. These covariance differences are within the uncertainty of the assumed applicability of linear error theory.

TABLE 6.4

ITERATIVE KALMAN FILTER (MODE 2) SOLUTION 6.1B

at t = 0 - (starting estimates)

$$\hat{X} = \begin{Bmatrix} -5000 \\ -1500 \\ 50 \\ 250 \\ 50 \\ 400 \end{Bmatrix}, P = \begin{bmatrix} 10^6 & & & & & 0 \\ & 10^6 & & & & \\ & & 10^6 & & & \\ & & & 10^6 & & \\ & & & & 10^6 & \\ 0 & & & & & 10^6 \end{bmatrix}$$

at t = 0 + (after first observation subset processed)

$$\hat{X} = \begin{Bmatrix} -10018 \\ -2954 \\ 79 \\ 250 \\ 50 \\ 400 \end{Bmatrix}, P = \begin{bmatrix} 2512 & -61 & 2 & 2 & 0 & 0 \\ -61 & 2699 & 0 & 0 & 0 & 0 \\ 2 & 0 & 2719 & 0 & 0 & 0 \\ 0 & 0 & 0 & 10^6 & 0 & 0 \\ 0 & 0 & 0 & 0 & 10^6 & 0 \\ 0 & 0 & 0 & 0 & 0 & 10^6 \end{bmatrix}$$

at t = 38 (at the end of the first *forward* pass through the data)

$$\hat{X} = \begin{Bmatrix} 9019 \\ 798 \\ 7284 \\ 501 \\ 100 \\ -423 \end{Bmatrix}, P = \begin{bmatrix} 607 & -96 & -211 & 29 & -3 & -6 \\ -96 & 511 & 38 & -4 & 25 & 2 \\ -211 & 38 & 743 & -9 & 2 & 30 \\ 29 & -4 & -9 & 1.5 & -0.1 & -0.2 \\ -3 & 25 & 2 & -0.1 & 1.4 & 0.1 \\ -6 & 2 & 30 & -0.2 & 0.1 & 1.6 \end{bmatrix}$$

at t = 0 (at the end of the first *reverse* pass)

$$\hat{X} = \begin{Bmatrix} -10020 \\ -2986 \\ 164 \\ 500.9 \\ 99.6 \\ 798.4 \end{Bmatrix}, P = \begin{bmatrix} 584 & -15 & 57 & -28 & 2 & 2 \\ -15 & 756 & 90 & 2 & -33 & -4 \\ 57 & 90 & 961 & 0 & -4 & -38 \\ -28 & 2 & 0 & 1.5 & -.2 & -.2 \\ 2 & -33 & -4 & -.2 & 1.6 & 0.2 \\ 2 & -4 & -38 & -.2 & 0.2 & 1.9 \end{bmatrix}$$

at t = 0 (at the end of the second *reverse* pass)

$\hat{X}$, P unchanged from above values

Rigorous proof that a Mode 2 iterative application of the extended sequential Kalman formulation always converges to the same estimates as the batch least squares formulations is not availiable in the literature. While the two approaches can be viewed as being based upon the same general principle that the model parameters be "adjusted" to bring the predicted and observed behavior into least square agreement, it must be recognized that the implementation of sequential and batch estimation algorithms make use of linearizations in a different fashion. Traditionally, the linear error theory covariance matrix of the batch algorithm is "believed" only after final convergence is achieved. However, the extended Kalman algorithm makes use of the intermediate state covariance estimates to "weight" each intermediate state estimate. While one contrived example has shown {Fang (1975)} that convergence to different estimates can occur, the author does not know of a practical application which demonstrates this convergence anomaly. In fact, Shutz (1973) found that the Mode 2 algorithm (applied to orbit determination) reliably converges to the batch least squares estimates; he also found the radius of convergence to be about a factor of ten greater than the batch algorithm.

6.2 *Loss of Precision in Covariance Propagation Algorithms*

It is a well-known truth that the classical Kalman covariance recursion equation

$$P_{k+1} = P_k - P_k A_{k+1}^t (\Lambda_{v_{k+1}v_{k+1}} + A_{k+1} P_k A_{k+1}^t)^{-1} A_{k+1} P_k \quad (6.28)$$

{developed in § 1.4 (see eqn. (1.93)) and the extensions (eqns. (5.42b) and (5.86) for dynamical systems} often suffers from severe loss of precision due to accumulation of arithmetic errors. The loss of precision is occasionally so damaging that the elements of P contain no significant figures and P may lose its positive-definite property (a negative variance is not only incorrect, it is also impossible!)

Several approaches have been successfully used to circumvent this problem:

(1) Increase the number of significant figures in the covariance recursion calculations with equation (6.28), (5.42b) or (5.86).

(2) Develop alternative covariance recursions which are more stable numerically.

(3) Introduce artificial "bounding" logic which resets elements of P if prescribed bounds are violated.

The first two approaches (or a combination thereof) are much preferred to the third (indeed, it is difficult to defend the third procedure). A recent paper by Thornton and Bierman (1977) illustrates the presence of the above problem and demonstrates the relative merits of several algorithms for overcoming this numerical difficulty. After a brief discussion of matrix factorization covariance update algorithms, we will summarize Thornton and Bierman's results and major conclusions.

6.2.1 Matrix Factorization Covariance Update Algorithms

As developed in Appendix C, any positive definite matrix P can be factored as

$$P = SS^t \quad (\text{"square root factorization"}) \tag{6.29}$$

or as

$$P = UDU^t \quad (\text{"U-D factorization"}) \tag{6.30}$$

S is an upper or lower triangular "square root matrix", if determined numerically from P, S can be calculated using the Cholesky decomposition algorithm (C.57). U and D are unit upper triangular and diagonal matrices, respectively.

In particular, if the covariance matrix P is initially factored according to (6.29) or (6.30), one can consider the wisdom of replacing (6.28) by recursion equations for updating the factors S or U and D. Carlson (1973) and Kaminski, et al (1971) have developed and applied algorithms for updating S, while Thornton and Bierman (1975, 1977) have developed and applied algorithms for updating U and D. Thornton and Bierman's study establish that both the square root and U-D covariance update algorithms are less prone to numerical difficulty than equation (6.28) {in fact, single precision calculations with the square root or U-D update equations are usually competitive with double precision calculations using eqns. (6.28)}. The U-D update algorithm is more efficient than is the SS^t update algorithm. Accordingly, we summarize only the U-D covariance recursion algorithm. We will, however, compare numerical results using the conventional algorithm, the U-D algorithm and one implementation of the square root algorithm (the Potter-Schmidt algorithm); the interested reader is encouraged to consult the cited papers for further details.

For notational compactness, denote the apriori covariance as $P \equiv P_k$ and the updated covariance matrix as $\hat{P} \equiv P_{k+1}$. Let the U-D factors of P and $\hat{P}$ be similarily denoted as

$$P \equiv U\,D\,U^t \tag{6.31}$$

and

$$\hat{P} = \hat{U}\,\hat{D}\,\hat{U}^t\ . \tag{6.32}$$

We now summarize Bierman's (1975) algorithm for updating the U-D factors of P and for calculating the Kalman gain matrix.

Given U, D, A, and σ^2, where A is a 1 x n constant coefficient matrix for the scalar measurement

$$\tilde{y} = A\,X + V \tag{6.33}$$

and

$$\sigma^2 = E(V^2),$$

determine $\hat{U}$, $\hat{D}$ and the Kalman gain matrix K by the following sequence of calculations:

$$f^t \equiv (f_1 \cdots f_n) = A^t U$$

$$z_i = d_i \, f_i \quad , \; i = 1,2,\cdots,n$$

$$\alpha_1 = \sigma^2 + z_1 \, f_1$$

$$\hat{d}_1 = d_1 \, \sigma^2/\alpha_1$$

$$b_1 = z_1$$

FOR $j = 1,2,\cdots,n$

$$\alpha_j = \alpha_{j-1} + f_j \, z_j$$
$$\hat{d}_j = d_j \, \alpha_{j-1}/\alpha_j \tag{6.34}$$

$$b_j \leftarrow z_j$$

$$p_j = -f_j/\alpha_{j-1}$$

$$\left.\begin{aligned} \hat{U}_{ij} &= U_{ij} + b_i \, p_j \\ b_i &\leftarrow b_i + U_{ij} \, z_j \end{aligned}\right\} \quad i = 1,2,\cdots,j-1$$

NEXT j

$$K = \frac{1}{2n} \, b$$

6.2.2 Thornton and Bierman's Case Study

The Mariner-Jupiter-Saturn (MJS '77) deep space mission was used by Thornton and Bierman (1977) to do a case study of several covariance recursion equations; we will summarize some of their results. The portion of the orbit state history being estimated is from 30 days prior to encounter of Saturn until encounter (passage through point of closest approach). Simulated measurements of spacecraft motion include range-rate measurements ($1\text{mm}^2/\text{second}^2$ variance) and an occasional range (distance) measurement (9 meter^2 variance); a total of 535 range rate and 72 range measurements were simulated to span the 30 day period. The state vector estimated included 3 position, 3 velocity and 3 acceleration variables (estimation of acceleration allows estimation of and compensation for unmodeled accelerations). All of these state variables were estimated as small departures from their respective values along a nominal orbit to facilitate linearizations. In addition to the nine state variables, one dynamic model parameter (Saturn's mass) and nine measurement model parameters (the three coordinates of each of three tracking stations) were estimated.

The conventional Kalman formulation {eqns. (5.42)} was employed, the state vector had 19 elements:

$$X = \left\{ \begin{array}{l} \text{3 position coordinates} \\ \text{3 velocity coordinates} \\ \text{3 acceleration coordinates} \\ \text{9 tracking station coordinates} \\ \text{mass of Saturn} \end{array} \right\}$$

The Kalman filter calculations were carried out in both single precision (SP) and double precision (DP); all calculations being performed using the Jet Propulsion Laboratory's UNIVAC 1108 computer*. In addition to the conventional Kalman filter calculations, modified Kalman filter algorithms were employed with three alternative covariance update algorithms.

(1) The U-D update algorithm presented in §6.2.1

(2) The Potter-Schmidt SS^t update algorithm {Kaminski, et at (1971)}

(3) The "stabilized Kalman" algorithm {Gentleman (1973)}

Figure 6.3 summarizes the performance of the conventional Kalman algorithm and the above three alternative algorithms for single (SP) and double precision (DP) calculations. The vertical axis is the estimated-minus-true position error (root-sum-square of the estimated-minus-true errors of the first three elements of X),

*Single precision (SP) on the UNIVAC 1108 corresponds to 8 significant digits.

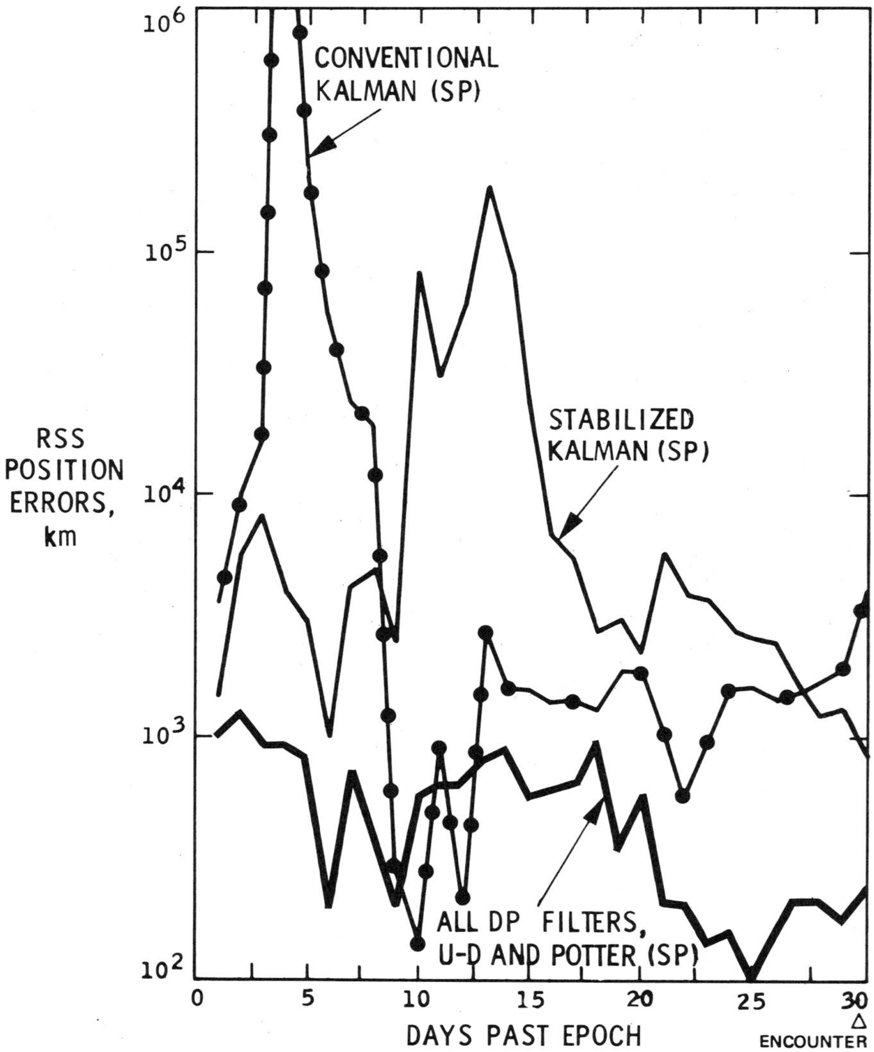

Figure 6.3 Comparison of Covariance Propagation Algorithms [reproduced by permission, Thornton and Bierman (1977)]

the horizontal axis is time. (The discussion is restricted to position errors for simplicity, analogous results were obtained for all 19 elements of X).

All four covariance update algorithms gave identical results when double precision was employed (the bold line of Figure 6.3). However, both the conventional Kalman update algorithm and the stabilized Kalman update algorithms produced erroneous estimates when the calculations were performed in single precision. Both of the matrix factorization covariance propagation algorithms, however produced accurate results using only single precision.

Table 6.1 summarizes the execution times for the four algorithms compared in Figure 6.3.

Table 6.1 UNIVAC 1108 Computer Execution Time (Seconds)

Algorithm employed to update covariance	Single-Precision	Double-Precision
Conventional Kalman	39	49
Stabilized Kalman	45	59
U-D(Thornton-Bierman)	38	46
SS^t(Potter-Schmidt)	63	80

These comparisons generally favor the U-D matrix factorization approach, both from accuracy and efficiency viewpoints. The present discussion represents a significant portion of the extensive study by Thornton and Bierman (1977) which supply additional evidence supporting the conclusion that the U-D factorization update algorithm (6.34) is superior to all known methods for propagating the covariance matrix.

6.3 Dynamically Constrained Satellite Photogrammetry

As a final example application of the Gaussian least square differential correction algorithm (§1.5), we will augment Example

Problem 3.4. With reference to figures 3.9-3.11, we previously considered each photograph's position and orientation to be unknown; thus six independent unknowns were required to establish the position (X_{cj}, Y_{cj}, Z_{cj}) and orientation (ϕ_{1j}, ϕ_{2j}, ϕ_{3j}) of the jth photograph. Clearly, a strip of 100 photographs would thus require determination of 600 unknowns to specify fully the exposure positions and orientations. The rather intimidating dimensionality of this situation motivates the consideration of alternatives; it is possible to reduce the number of unknown parameters dramatically through introduction of *dynamical constraints*.

The colinearity equations (3.29) models the location in the jth photograph of the ith measurable point via a functional relationship of the form

$$x_{pij} = F(X_{pi},Y_{pi},Z_{pi},X_{cj},Y_{cj},Z_{cj},\phi_{1j},\phi_{2j},\phi_{3j})$$
$$y_{pij} = G(X_{pi},Y_{pi},Z_{pi},X_{cj},Y_{cj},Z_{cj},\phi_{1j},\phi_{2j},\phi_{3j}) \qquad (6.35)$$

This model is based solely upon geometric optics.

As is demonstrated in Example Problem 3.4, all of the subscripted arguments on the right side of (6.35) can be recovered by least square differential correction, treating all of the parameters as *independent* unknowns. This is the traditional approach to the problem. The introduction of orbital dynamical constraints, conceived by Brown and implemented by Light, allows the elimination of the 3n (n= number of photographs) camera center coordinates in favor of 6 osculating orbital elements. Blanton and Junkins (1976) carried the incorporation of dynamical constraints to its logical conclusion through rigorous satisfaction of the satellite equations of rotational motion. This process allows the further reduction of unknowns from 3n orientation angles to 6 osculating attitude elements or constants of the rotational motion. The osculating attitude elements chosen are the initial angular velocities and 1-2-3 Euler angles of the motion.

Orbit Dynamical Constraints

The essence of orbit constrained photogrammetry is the recognition that the camera exposure stations along a given strip of n photographs are dynamically constrained according to

$$
\begin{aligned}
X_{c_j} &= X_c(t_j, c_1, c_2, \cdots, c_6) \\
Y_{c_j} &= Y_c(t_j, c_1, c_2, \cdots, c_6) \quad j = 1, 2, \cdots, n \\
Z_{c_j} &= Z_c(t_j, c_1, c_2, \cdots, c_6)
\end{aligned}
\tag{6.36}
$$

where Eqns. (6.36) are functional representations of the solution of the spacecraft's translational equations of motion:

$$
\begin{aligned}
\ddot{X}_c &= -GMX_c/r^3 + \text{perturbations} \\
\ddot{Y}_c &= -GMY_c/r^3 + \text{perturbations} \\
\ddot{Z}_c &= -GMZ_c/r^3 + \text{perturbations} \\
r^2 &= X_c^2 + Y_c^2 + Z_c^2
\end{aligned}
\tag{6.37}
$$

The six constants (c_1, $c_2, \cdots c_6$) can be any set of initial conditions or osculating orbital elements which uniquely define the solution of (6.37). The least squares solution of the photogrammetry problem is modified to include the dynamical constraints as follows: Current estimates of c_1, $c_2, \cdots$, c_6 are used in Equations (6.37) to compute $X_{c_j}, Y_{c_j}, Z_{c_j}$ at each photograph exposure time t_j. These resulting current estimates of camera exposure coordinates are used in Equations (6.35) together with the other parameter estimates $(X_{p_i}, Y_{p_i}, Z_{p_i}, \phi_{1j}, \phi_{2j}, \phi_{3j})$ to determine current computed values of $x_{p_{ij}}, y_{p_{ij}}$ which are in turn differenced from the corresponding observed values to find the current residual vector. The partial derivatives of the observation Equations (6.35) with respect to the orbital elements, $c_1, c_2, \cdots, c_6$, (the elements of the "A" matrix) are determined by the chain differentiation rule applied to Equations (6.35) and (6.36). For example,

$$\frac{\partial x_{p_{ij}}}{\partial c_1} = \frac{\partial F}{\partial X_{c_j}} \frac{\partial X_{c_j}}{\partial c_1} + \frac{\partial F}{\partial Y_{c_j}} \frac{\partial Y_{c_j}}{\partial c_1} + \frac{\partial F}{\partial Z_{c_j}} \frac{\partial Z_{c_j}}{\partial c_1} .$$

The partial derivatives with respect to the other parameters are identical in form to the corresponding equations for unconstrained photogrammetry. The partial derivatives of the solution (6.36) of (6.37) are determined using the methods of §4.5.

Satellite Rotational Dynamical Constraints

If the satellite is idealized as a near-rigid body, the rotational motion is governed by the system of nonlinear differential equations {see reference 6.12, for example}.

$$\begin{Bmatrix} \dot{\phi}_1 \\ \dot{\phi}_2 \\ \dot{\phi}_3 \end{Bmatrix} = \frac{1}{\cos\phi_2} \begin{pmatrix} \cos\phi_3 & -\sin\phi_3 & 0 \\ \cos\phi_2 \sin\phi_3 & \cos\phi_2 \cos\phi_3 & 0 \\ -\sin\phi_2 \cos\phi_3 & \sin\phi_2 \sin\phi_3 & \cos\phi_2 \end{pmatrix} \begin{Bmatrix} \omega_1 \\ \omega_2 \\ \omega_3 \end{Bmatrix} \tag{6.38a}$$

$$\begin{aligned} I_1\dot{\omega}_1 &= (I_2-I_3)\omega_2\omega_3 + L_1(t,\phi_1,\phi_2,\phi_3,\dot{\phi}_1,\dot{\phi}_2,\dot{\phi}_3,X_c,Y_c,Z_c) \\ I_2\dot{\omega}_2 &= (I_3-I_1)\omega_1\omega_3 + L_2(t,\phi_1,\phi_2,\phi_3,\dot{\phi}_1,\dot{\phi}_2,\dot{\phi}_3,X_c,Y_c,Z_c) \\ I_3\dot{\omega}_3 &= (I_1-I_2)\omega_1\omega_2 + L_3(t,\phi_1,\phi_2,\phi_3,\dot{\phi}_1,\dot{\phi}_2,\dot{\phi}_3,X_c,Y_c,Z_c) \end{aligned} \tag{6.38b}$$

where

$\{\phi_1,\phi_2,\phi_3\} \equiv$ 1-2-3 Euler angles orienting the camera axes (spacecraft-fixed) in inertial space.

$\{\omega_1,\omega_2,\omega_3\} \equiv$ body principal axes components of angular velocity (for simplicity, the body axes and camera axes are assumed coincident)

$\{I_1,I_2,I_3\} \equiv$ principal moments of inertia

$\{L_1,L_2,L_3\} \equiv$ body axis components of the instantaneous torque acting on the satellite.

The solution of equations (6.38), carried out numerically using the methods of chapter 4 (or analytically for special cases) has the functional structure

$$\phi_{ij} \equiv \phi_i(t_j) = \phi_i\,(t_j - t_o, c_7, c_8, \cdots, c_{12}, I_1, I_2, I_3)$$

$$i = 1,2,3 \tag{6.39}$$

The six constants $(c_7, c_8, \cdots, c_{12})$ may be any set of initial conditions (on $\phi_1, \phi_2, \phi_3, \dot{\phi}_1, \dot{\phi}_2, \dot{\phi}_3$) or other constants which uniquely defines the solution of equations (6.38). The inclusion of rotational dynamical constraints is analogous to including translational (orbital) dynamical constraints. Current estimates of $(c_7, c_8, \cdots, c_{12})$ are used to solve equations (6.38) in the form (6.39) at each exposure time t_j. The modified least square triangulation problem in which the unknowns $\{\phi_{1j}, \phi_{2j}, \phi_{3j}\}$ $j=1,2,\cdots$ have been eliminated in favor of the six constants $(c_7, c_8, \cdots, c_{12})\}$ now requires partial derivatives of equations (6.35) with respect to the c's as , for example,

$$\frac{\partial X_{pij}}{\partial c_k} = \frac{\partial F}{\partial c_k} = \frac{\partial F}{\partial \phi_{ij}} \frac{\partial \phi_{ij}}{\partial c_k} + \frac{\partial F}{\partial \phi_{2j}} \frac{\partial \phi_{2j}}{\partial c_k} + \frac{\partial F}{\partial \phi_{3j}} \frac{\partial \phi_{3j}}{\partial c_k} \qquad k = 7,8,\cdots,12$$

The partial derivatives $\frac{\partial \phi_{ij}}{\partial c_k}$ of the solution (6.39) of (6.38) are determined numerically (in general), using the methods of §4.5.

This discussion can be focused by consideration of the impact of dynamical constraints upon the dimensionality of a typical application. Table 6.2 summarizes the number of measurements and number of unknowns for triangulating a strip of 30 photographs. As is evident, the number of unknowns can be reduced from 480 to 312 through incorporation of translational (orbital) and rotational (attitude) dynamical constraints.

This example, however, is still of rather intimidating size for test calculations. We therefore consider a very small strip

Table 6.2 A TYPICAL PHOTOGRAMMETRIC TRIANGULATION PROBLEM

30	photographs (single strip)
100	pass points
50	control points
1200	measurements ($\approx$ 20 measured points/photo)

WITH NO DYNAMICAL CONSTRAINTS:

90	camera position coordinates
90	camera orientation coordinates
300	pass point coordinates
480	total # of unknowns

WITH TRANSLATIONAL DYNAMICAL CONSTRAINTS:

6	orbit initial conditions
90	camera orientation coordinates
300	pass point coordinates
396	total # of unknowns

WITH TRANSLATIONAL & ROTATIONAL DYNAMICAL CONSTRAINTS:

6	orbit initial conditions
6	attitude initial conditions
300	pass point orientation coordinates
312	total # of unknowns

Table 6.3 PHOTOGRAMMETRIC TRIANGULATION TEST PROBLEM 6.3

4	photographs
4	control points
8	pass points
60	measurements

WITH NO DYNAMICAL CONSTRAINTS:

12	camera position coordinates
12	camera orientation coordinates
24	pass point coordinates
48	total unknowns

WITH ORBIT DYNAMICAL CONSTRAINTS:

6	orbit initial conditions
12	camera orientation coordinates
24	pass point coordinates
42	total unknowns

WITH ORBIT AND ATTITUDE DYNAMICAL CONSTRAINTS:

6	orbit initial conditions
6	attitude initial conditions
24	pass point coordinates
36	total unknowns

of only four photographs made by a camera fixed in a rigid triaxial satellite; placed in 100 mile circular polar orbit over a non-rotating earth. Two-body translational motion and torque-free rotational motion were assumed. A single strip of four photographs, each with approximately 50% overlap was generated. These photographs covered a ground strip of approximately 100 miles by 250 miles.

Four control points and eight pass points were chosen on the ground strip and distributed in such a fashion as to give a total of 60 observation equations. The observation coordinates were modified to simulate 1 mil measurement precision. It was further assumed that the camera focal length, f, the principal point offset (x_0, y_0), and the camera exposure times were perfectly known. Initial guesses of the pass point coordinates and camera center coordinates were corrupted from their true values by an average of 2 miles. Initial guesses on camera axes Euler angles were corrupted by an average of 1.5°.

Three separate least squares differential correction programs were written. All computer runs were made on the University of Virginia CDC 6400 computer. These programs are summarized as follows:

UCPHOTO (unconstrained photogrammetry): all parameters appearing in the observation equations were treated as independent. The dimensions of the A matrix* were 60 (observations) x 48 (parameters), where 48 = 3 times the number of pass points plus 3 times the number of camera center locations plus 3 camera axes Euler angles at each exposure time = 3(8) + 3(4) + 3(4).

OCPHOTO (orbit constrained photogrammetry): pass point coordinates, camera orientation at each time and orbit initial conditions were considered independent. The A matrix was dimensioned 60

$${}^{*}A = \left\{\frac{\partial\ (\text{MEASUREMENTS})}{\partial\ (\text{UNKNOWN PARAMETERS})}\right\}$$

(observations) x 42 (parameters), where 42 = 3 times the number of pass points plus 6 orbit initial conditions plus 3 Euler angles for each photograph = 3(8) + 6 + 3(4).

DCPHOTO (fully dynamically constrained photogrammetry): pass point coordinates, orbit initial conditions and attitude initial conditions were considered independent. The dimensions of the A matrix were 60 (observations) x 36 (parameters) where 36 = 3 times the number of pass points plus 6 orbit initial conditions plus 6 attitude initial conditions = 3(8) + 6 + 6.

The structure of the A matrix for these three cases is shown in Figures 6.4a, 6.4b, and 6.4c, respectively. The shaded region contains the non-zero elements. The essential results of these numerical tests are summarized in Table 6.4. For this simple problem the best results were obtained using full dynamical constraints. (Even though the dimension of the A matrix was reduced from 60 x 48 to 60 x 36, the computer run time was not decreased). This is because the matrix inversion savings were offset by the relatively elaborate dynamical model. The matrix calculations, however, dominate real world triangulation problems. Consequently, substantial cost reductions may be achieved in practical applications.

Program	Max pass pt error after convergence (true - converged) (ft)	Mean pass point error (ft)	Std. dev. of all errors (ft)	CDC 6400 central processor time (sec)
UCPHOTO	205.08	12.24	73.58	5.82
OCPHOTO	93.45	-6.36	37.47	5.66
DCPHOTO	-57.88	-1.60	25.54	6.86

Table 6.4 A Comparison of the effects of dynamical constraints on photogrammetric triangulation.

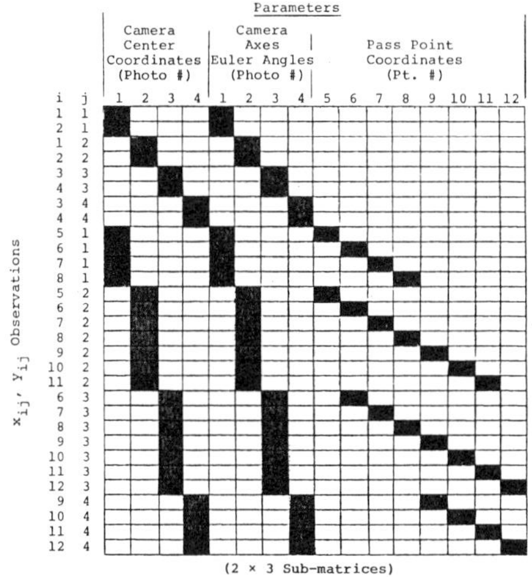

Fig. 6.3a No Dynamical Constraints

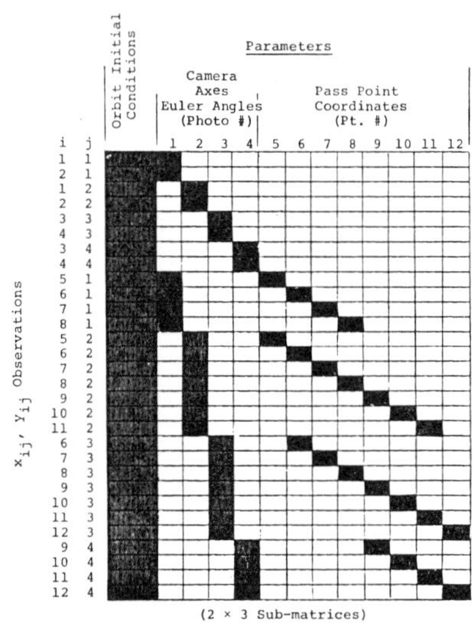

Fig. 6.3b Orbit Dynamical Constraints

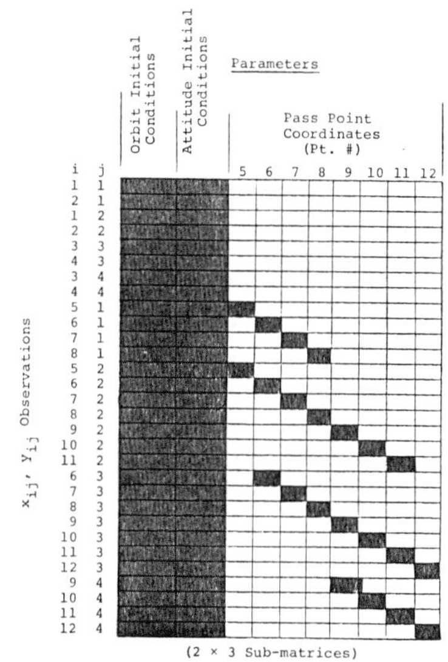

Fig. 6.3c Fully Dynamical Constraints

Figure 6.3 Structure of the A matrix $\left(\dfrac{\partial(\text{measurements})}{\partial(\text{unknown parameters})}\right)$ for Unconstrained, Orbit constrained and Attitude constrained Triangulation of Orbital Photography

Concluding Remarks on the Dynamically Constrained Triangulation Problem

The rotational dynamics model underlying the application is based on the assumption that the motion is governed by the torque-free form of Euler's equations. This is in general a restrictive assumption, although there may be applications for which this model is adequate. Even in the presence of small torques, an osculating torque-free solution often provides good results over small elapsed times.

From the results summarized in Table 6.2, we conclude that the application of full dynamical constraints to photogrammetric triangulation appears feasible. The problem of bookkeeping associated with large real life problems has not been addressed here but would essentially be unchanged by the inclusion of dynamical constraints. It is also impossible, at this point, to make any quantitative statements about a potential reduction of computer execution time for a large problem. The various tradeoffs, primarily between more computation required for the inclusion of dynamical constraints, and the resulting less computation for the solution of the lower dimensioned least squares normal equations needs further study. The obvious advantages of the inclusion of dynamical constraints (given an adequate dynamic model of the motion) are the increased observability of the system and the numerical benefits ensuing from reduction in the number of unknowns in the problem. These advantages are evident even in the small problem discussed herein. In a very large problem, a solution might not otherwise be obtainable. The description of the computer implementation details and documentation of numerical examples were rather constrained in the present discussion. The interested reader is referred to Blanton (1976). Background discussions on the photograph triangulation problem are found in Light (1972).

6.4 Remarks

The first example (§6.1) of this chapter was developed during a short course taught to the staff of the Naval Surface Weapons Center, Dahlgren, Va. (1974). The transparent dynamical model and measurement geometry of this example make it an especially attractive basis for studying estimation algorithms. The applications of §6.2 and §6.3 are more involved; it was necessary to summarize essential results and to reference papers which contain in depth discussions which interested readers may pursue. These applications, are most instructive, since they embody several hard-won insights likely to prove useful applications of the estimation methodology developed in this text.

6.5 References for Chapter 6

6.1 Shutz, B.E., J.D. McMillan, and B.D. Tapley, "A Comparison of Estimation Methods for the Reduction of Laser Observations of a Near-Earth Satellite", presented to the AAS/AIAA Astrodynamics Conference, Vail, Colorado, July 1973.

6.2 Fang, B.T., "Nonlinear Counter Example for Batch and Extended Sequential Estimation Algorithms", Flight Mechanics/Estimation Theory Symposium, NASA GSFC preprint X-582-75-273, Greenbelt, Md., August 1975.

6.3 Thornton, C.L., and G.J. Bierman, "A Numerical Comparison of Discrete Kalman Filtering Algorithms: An Orbit Determination Case Study", JPL Technical Memorandum 33-771, to appear in Automatica, Jan. 1977.

6.4 Carlson, N.A., "Fast Triangular Formulation of the Square Root Filter", AIAA Journal, Vol. 11, No. 9, 1973, pp 1259-1264.

6.5 Kaminski, P.G., A.E. Bryson, and S.F. Schmidt, "Discrete Square Root Filtering: A Survey of Current Techniques", IEEE Trans. on Automatic Control, Vol. AC-16, No. 6, pp 727-736, 1971.

6.6 Bierman, C.J., "Measurement Updating Using the U-D Factorization", Proceedings 1975 IEEE Conference on Decision and Control, pp 337-346, 1975.

6.7 Gentleman, W.M., "Least Square Computations by Givens Transformations Without Square Roots", J. Inst. Maths. Applic., Vol. 12, pp 329-336, 1973.

6.8 Brown, D. "The Rigorous and Simultaneous Adjustment of Lunar Orbital Photography, Rome Air Development Center Report RADC-TR-70-274, 1970.

6.9 Light, D.L., "Photo Geodesy from Apollo", Photogrammetric Engineering, Vol. 38, No. 6, pp 547-587, 1972.

6.10 Junkins, J.L., "Estimation Theory", Naval Surface Weapons Center Report No. TN-K-45h4, Dahlgren, Va., January 1975.

6.11 Blanton, J.N., and J.L. Junkins, "Dynamical Constraints in Satellite Photogrammetry", AIAA Paper #76-824, AIAA/AAS Astrodynamics Conference, San Diego, Ca., August 1976, to appear in AIAA Journal, May 1977.

6.12 Blanton, J.N., "A New Formulation of the Free Motion of Triaxial Rigid Bodies and Applications to Satellite Photogrammetric Triangulation", Ph.D. Dissertation, Univ. of Va. Charlottesville, Va., August 1976.

APPENDIX A

Minimization of Functions of N Variables

In this appendix, classical necessary and sufficient conditions for solution of unconstrained and equality-constrained parameter optimization problems are summarized. We also summarize two iterative techniques for unconstrained minimization, and discuss the relative merits of these approaches.

A.1 Unconstrained Extrema

Table A1 summarizes the classical necessary and sufficient conditions which must be satisfied for a function $\phi = f(x_1,\ldots,x_n)$ to have a maximum, minimum, or indefinite (saddle) point. The usual approach is to attempt to determine all "stationary" or "critical" points satisfying the necessary conditions and then employ the sufficient conditions to test whether each such point $(\hat{x}_1,\ldots,\hat{x}_n)$ is a maximum or minimum of ϕ. The sufficiency condition requires that one determine the *definiteness* of the matrix of second partial derivatives, therefore we digress to consider this concept.

Let: A = A given n x n matrix.

X = An arbitrary non-zero n x 1 vector.

If	Then matrix A is
$X^tAX > 0$	positive definite
$X^tAX < 0$	negative definite
$X^tAX \geq 0$	positive semi-definite
$X^tAX \leq 0$	negative semi-definite
$X^tAX \gtreqless 0$	indefinite

Definiteness of a given matrix can be determined (although not

TABLE A1

UNCONSTRAINED PARAMETER OPTIMIZATION

(Summary of Necessary and Sufficient Conditions)

1 variable	n variables
<u>Problem</u>: Determine $x = \hat{x}$ such that $\phi = \phi(x)$ is a local max. or min.	<u>Problem</u>: Determine $\{\hat{x}_1, \ldots, \hat{x}_n\}$ such that $\phi = \phi(x_1, x_2, \ldots, x_n)$ is a local maximum or minimum.
<u>Necessary Conditions</u>: $\left.\frac{d\phi}{dx}\right\rvert_{x=\hat{x}} = 0$ (solve for all $\hat{x}$'s "stationary points")	<u>Necessary Conditions</u>: $\left.\frac{\partial\phi}{\partial x_i}\right\rvert_{\wedge} = 0 \quad i = 1, 2, \ldots, n$ n equations ... solve for all $\{\hat{x}_1, \ldots, \hat{x}_n\}$ which satisfy them
<u>Sufficient Conditions</u>: $\left.\frac{d^2\phi}{dx^2}\right\rvert_{x=\hat{x}} \begin{cases} > 0 & \text{local min.} \\ = 0 & \text{inflection} \\ < 0 & \text{local max.} \end{cases}$	<u>Sufficient Conditions</u>: ① $\sum_{i=1}^{n} \sum_{j=1}^{n} \left.\frac{\partial^2\phi}{\partial x_i \partial x_j}\right\rvert_{\wedge} \Delta x_i \, \Delta x_j \equiv q$ $q \begin{cases} > 0 & \text{local min.} \\ \lessgtr 0 & \text{saddle pt.} \\ < 0 & \text{local max.} \end{cases}$ for all $\{\Delta x_i, \Delta x_j\}$: arbitrary displacements away from $\{\hat{x}_1, \hat{x}_2, \ldots, \hat{x}_n\}$. The double sum ① can be written in matrix form as ② $\Delta x^t \{\nabla_x^2 \phi\} \, \Delta x \begin{cases} > 0 & \text{minimum} \\ \gtrless 0 & \text{saddle} \\ < 0 & \text{maximum} \end{cases}$ Equation ② requires that the matrix of second partials $\{\nabla_x^2 f\} = \left(\left.\frac{\partial^2\phi}{\partial x_i \partial x_j}\right\rvert_{\wedge}\right)$ to be *positive definite* (local minimum), *negative definite* (local maximum), or *indefinite*.

without occasional numerical difficulties) via one of the following tests:

Test 1 (Sylvester's Criterion)

Let $A = \begin{bmatrix} a_{11} & \cdots & a_{1n} \\ \vdots & \ddots & \vdots \\ a_{n1} & \cdots & a_{nn} \end{bmatrix}$

$$D_2 = \det \begin{bmatrix} a_{11} & a_{12} \\ a_{21} & a_{22} \end{bmatrix}$$

$$\vdots$$

$$D_i = \det \begin{bmatrix} a_{11} & \cdots & a_{1i} \\ \vdots & & \\ a_{i1} & \cdots & a_{ii} \end{bmatrix}$$

$$D_n = \det A$$

Then necessary and sufficient conditions for definiteness are:

A is positive definite if $D_i > 0$, $i=1,2,\ldots,n$

A is negative definite if $D_i > 0$, $i=2,4,\ldots,i$ even; $D_i < 0$, $i=1,3,\ldots,i$ odd

Otherwise A is indefinite .

Test 2 (Eigenvalue Test)

Let the *characteristic polynomial* be defined by

$$\det \{A-\lambda I\} = 0 \tag{A.1}$$

and let the n roots of (A.1) be the eigenvalues $(\lambda_1,\lambda_2,\ldots,\lambda_n)$.

If the eigenvalues are all real and positive, then A is positive definite.

If the eigenvalues are all real and negative, then A is negative definite.

Otherwise A is indefinite.

The proofs of these tests are given by Ogata (1967).

A.2 *Equality Constrained Extrema*

One often encounters problems of the form:

Extremize

$$\phi = \phi(x_1,\ldots,x_n) = \phi(X) \tag{A.2}$$

subject to

$$\psi_j(x_1,\ldots,x_n) = \psi_j(X) = 0 \quad j=1,2,\ldots,m,\ m<n \tag{A.3}$$

As is established in Bryson and Ho (1969), the necessary conditions for a stationary point of (A.2) subject to (A.3) is obtained by applying the *unconstrained* necessary conditions of §A.1 to the *augumented function*

$$\Phi = \phi + \sum_{j=1}^{m} \lambda_j \psi_j \tag{A.4}$$

where $\{\lambda_1,\ldots,\lambda_m\}$ are a set of *Lagrange multipliers*. Thus the <u>necessary conditions</u> are

$$\left.\begin{aligned} \frac{\partial\Phi}{\partial x_i} &= \frac{\partial\phi}{\partial x_i} + \sum_{j=1}^{m} \lambda_j \frac{\partial\psi_j}{\partial x_i} = 0 \qquad i=1,2,\ldots,n \\ \frac{\partial\Phi}{\partial\lambda_j} &= \psi_j(x_1,\ldots,x_n) = 0 \qquad j=1,2,\ldots,m \end{aligned}\right\} \tag{A.5}$$

The n+m equations (A.5) are solved for the n+m unknowns $\{\hat{x}_1,\hat{x}_2,\ldots,\hat{x}_n;\ \hat{\lambda}_1,\hat{\lambda}_2,\ldots,\hat{\lambda}_m\}$, which are the constrained stationary points. Equations (A.4) and (A.5) are the *equations defining the Lagrange multiplier rule.*

Table A.2 summarizes the necessary and sufficient conditions for equality constrained maxima and minima.

Consider the following example application of the results of Table A2.

TABLE A2

EQUALITY CONSTRAINED PARAMETER OPTIMIZATION

(Necessary & Sufficient Conditions for Equality Constrained Extrema)

2 variables, 1 constraint	n variables, m constraints, m<n
Problem:	Problem:
Extremize	Extremize
$\phi(y,z)$	$\phi(x\ ,x\ ,\ldots,x_m,x_{m+1},\ldots,x_n) \equiv \phi(Y,Z)$
Subject to $\psi(y,z) = 0$	Subject to $\Psi(Y,Z) = 0$
Augumented function:	Augmented function:
$\Phi \equiv \phi + \lambda\psi$	$\Phi \equiv \phi + \Lambda^t\Psi, \quad \Lambda^t \equiv \{\lambda_1,\ldots,\lambda_m\}$
Necessary Conditions	Necessary Conditions
$\Phi_{\hat{x}} \equiv \left.\frac{\partial\Phi}{\partial y}\right\|_{\wedge} = 0$	$\Phi_{\hat{y}} \equiv \begin{Bmatrix} \left.\frac{\partial\Phi}{\partial x_1}\right\|_{\wedge} \\ \vdots \\ \left.\frac{\partial\Phi}{\partial x_m}\right\|_{\wedge} \end{Bmatrix} = 0$
$\Phi_{\hat{z}} \equiv \left.\frac{\partial\Phi}{\partial z}\right\|_{\wedge} = 0$	$\Phi_{\hat{z}} \equiv \begin{Bmatrix} \left.\frac{\partial\Phi}{\partial x_{m+1}}\right\|_{\wedge} \\ \vdots \\ \left.\frac{\partial\Phi}{\partial x_n}\right\|_{\wedge} \end{Bmatrix} = 0$
$\psi(\hat{y},\hat{z}) = 0$	$\Psi^t(\hat{Y},\hat{Z}) = \{\psi_1(\hat{Y},\hat{Z}),\ldots,\psi_m(\hat{Y},\hat{Z})\} = 0$
Sufficient Conditions	Sufficient Conditions
$q \begin{cases} > 0 & \text{local minimum} \\ < 0 & \text{local maximum} \end{cases}$	$Q \begin{cases} \text{positive definite, local minimum} \\ \text{negative definite, local maximum} \end{cases}$
where $q \equiv \Phi_{\hat{y}\hat{y}}\left(\frac{\psi_{\hat{z}}}{\psi_{\hat{y}}}\right)^2 - 2\left(\frac{\psi_{\hat{z}}}{\psi_{\hat{y}}}\right)\Phi_{\hat{y}\hat{z}} + \Phi_{\hat{z}\hat{z}}$	where $Q \equiv \Psi_{\hat{z}}^t(\Psi_{\hat{y}}^t)^{-1}\Phi_{\hat{y}\hat{y}}(\Psi_{\hat{y}}^t)^{-1}\Psi_{\hat{z}} - \Phi_{\hat{z}\hat{y}}(\Psi_{\hat{y}})^{-1}\Psi_{\hat{z}} - \Psi_{\hat{z}}^t(\Psi_{\hat{y}}^t)^{-1}\Phi_{\hat{y}\hat{z}} + \Phi_{\hat{z}\hat{z}}$

Example A1

Extremize

$$\phi = 6 - \frac{y}{2} - \frac{z}{3} \quad \text{(a plane)}$$

subject to

$$\psi(y,z) = 9(y-4)^2 + 4(z-5)^2 - 36 = 0 \quad \text{(an elliptic cylinder)}$$

form the augmented function

$$\Phi = 6 - \frac{y}{2} - \frac{z}{3} + \lambda\{9(y-4)^2 + 4(z-5)^2 - 36\}$$

<u>necessary conditions</u>

$$\Phi_{\hat{y}} \equiv \left.\frac{\partial\Phi}{\partial y}\right|_n = -\tfrac{1}{2} + 18\lambda(y-4) = 0 \rightarrow \hat{y} = 4 + \frac{1}{36\lambda}$$

$$\Phi_{\hat{y}} \equiv \left.\frac{\partial\Phi}{\partial z}\right|_n = -\frac{1}{3} + 8\lambda(z-5) = 0 \rightarrow \hat{z} = 5 + \frac{1}{24\lambda}$$

$$\psi(\hat{y},\hat{z}) = 9(\hat{y}-4)^2 + 4(\hat{z}-5)^2 - 36 = 0 \rightarrow \frac{1}{\lambda} = \pm\, 36\sqrt{2}$$

therefore

$$(\hat{y},\hat{z},\hat{\phi})_{1,2} = (4 \pm \sqrt{2},\ 5 \pm \tfrac{3}{2}\sqrt{2},\ \tfrac{7}{3} \pm \sqrt{2})$$

<u>sufficient conditions</u>

$$q = \Phi_{\hat{y}\hat{y}}\left(\frac{\psi_{\hat{z}}}{\psi_{\hat{y}}}\right)^2 - 2\left(\frac{\psi_{\hat{z}}}{\psi_{\hat{y}}}\right)\Phi_{\hat{y}\hat{z}} + \Phi_{\hat{z}\hat{z}}$$

$$= \ldots\ldots$$

$$= 8\lambda\left\{\frac{(\hat{z}-5)^2}{(\hat{y}-4)^2} + 1\right\}$$

thus if

$$\lambda \gtrless 0 \quad \text{then } q \gtrless 0$$

therefore

$$(\hat{y},\hat{z},\hat{\phi}) = (4 + \sqrt{2},\ 5 + \tfrac{3}{2}\sqrt{2},\ \tfrac{7}{3} - \sqrt{2}) \text{ is a local min. of } \phi$$

$$(\hat{y},\hat{z},\hat{\phi}) = (4 - \sqrt{2},\ 5 - \tfrac{3}{2}\sqrt{2},\ \tfrac{7}{3} + \sqrt{2}) \text{ is a local max. of } \phi$$

TABLE A3

EQUIVALENCE OF SUMMATION AND MATRIX EXPRESSIONS

Summation Equation	Equivalent Matrix Equation
Linear Scalar Equation	
$y = \sum_{i=1}^{n} a_i x_i$	$y = AX$ where $A = \{a_1 \ldots a_n\}$, $X = \begin{Bmatrix} x_1 \\ \vdots \\ x_n \end{Bmatrix}$
System of Linear Equations	
$y_j = \sum_{i=1}^{n} a_{ji} x_i; \quad j=1,2,..m$	$Y = AX$ where $Y = \begin{Bmatrix} y_1 \\ \vdots \\ y_m \end{Bmatrix}$, $X = \begin{Bmatrix} x_1 \\ \vdots \\ x_n \end{Bmatrix}$ $A = \begin{bmatrix} a_{11} & \cdots & a_{1n} \\ \vdots & \ddots & \vdots \\ a_{m1} & \cdots & a_{mn} \end{bmatrix}$
Quadratic Scalar Equation	
$z = \sum_{i=1}^{n} a_i x_i + \sum_{i=1}^{n} \sum_{j=1}^{n} b_{ij} x_i x_j$	$z = AX + X^t BX$ where $X = \begin{Bmatrix} x_1 \\ \vdots \\ x_n \end{Bmatrix}$, $A = \{a_1 \ldots a_n\}$, $B = \begin{bmatrix} b_{11} & \cdots & b_{1n} \\ \vdots & \ddots & \vdots \\ b_{n1} & \cdots & b_{nn} \end{bmatrix}$

TABLE A4

PARTIAL DERIVATIVES OF LINEAR AND QUADRATIC FORMS

Scalar Special Case	Matrix Generalization
Linear Equation	Linear Equation
$y = cx$	$\underset{1\times1}{y} = \underset{(1\times n)}{C^t}\ \underset{(n\times1)}{X} = c_1x_1 + \ldots + c_nx_n$
First Derivative	Gradient
$\frac{dy}{dx} = c$	$\nabla_x y \equiv \begin{Bmatrix} \frac{\partial y}{\partial x_1} \\ \vdots \\ \frac{\partial y}{\partial x_n} \end{Bmatrix} = C$
Quadratic Equation	Quadratic Form, B Symmetric
$y = ax + bx^2$	$y = \underset{(1\times n)}{A^t}\ \underset{(n\times1)}{X} + \underset{(1\times n)}{X^t}\ \underset{(n\times n)}{B}\ \underset{(n\times1)}{X}$
First Derivative	Gradient
$\frac{dy}{dx} = a + 2bx$	$\nabla_x y = A + 2BX$
Second Derivative	Matrix of Second Derivatives (Hessian)
$\frac{d^2y}{dx^2} = 2b$	$\nabla_x^2 y = \underset{n\times n}{\left[\frac{\partial^2 y}{\partial x_i \partial x_j}\right]} = 2B$
a,b,c constants, functions which do not depend upon x.	A,B,C matrices which do not depend upon $\{x_1, \ldots, x_n\}$.

Since all stationary points were investigated, these are the global constrained extrema.

A.3 Index vs Matrix Notations

Table A3 summarizes linear and quadratic algebraic expressions written in equivalent index/summation and matrix product forms. These equivalences, if not readily apparent, should be verified by carrying out the implied matrix algebra.

Table A4 summarizes some useful matrix calculus analogies which are often employed when the derivatives of linear and quadratic matrix products are required. Again, the reader should verify these results by taking partial derivatives of the matrix products, as carried out in terms of the matrix elements.

A.4 Nonlinear Unconstrained Optimization

A.4.1 Some Geometrical Insights

Consider the function $\phi(x,y)$ of two variables whose "contours" (lines of constant ϕ) are sketched in the figure A1

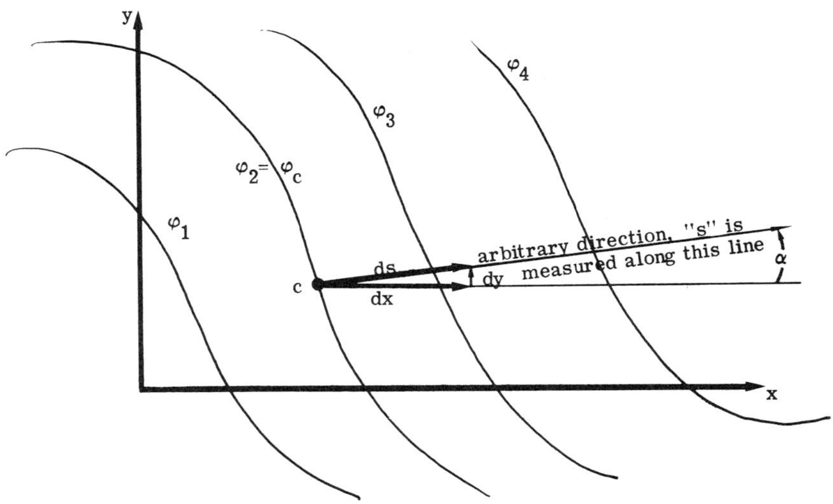

Figure A1 The Directional Derivative Concept

From the geometry of figure A1 it is evident that

$$\tan \alpha = \frac{dy}{dx} \tag{A.6a}$$

$$\frac{dy}{ds} = \sin \alpha \tag{A.6b}$$

$$\frac{dx}{ds} = \cos \alpha \ . \tag{A.6c}$$

For arbitrary small displacements (dx,dy) away from the "current" point (x_c, y_c), the differential change in ϕ is given by

$$d\phi = \left(\frac{\partial\phi}{\partial x}\Big|_c\right)dx + \left(\frac{\partial\phi}{\partial y}\Big|_c\right)dy. \tag{A.7}$$

If s is distance measured along an arbitrary line through c, then <u>the rate of change</u> ("directional derivative") of ϕ <u>in the direction of the line</u> is

$$\frac{d\phi}{ds}\Big|_c = \left(\frac{\partial\phi}{\partial x}\Big|_c\right)\left(\frac{dx}{ds}\Big|_c\right) + \left(\frac{\partial\phi}{\partial y}\Big|_c\right)\left(\frac{dy}{ds}\Big|_c\right) \tag{A.8}$$

and, making use of (A.6b) and (A.6c), we have

$$\frac{d\phi}{ds}\Big|_c = \left(\frac{\partial\phi}{\partial x}\Big|_c\right)\cos \alpha + \left(\frac{\partial\phi}{\partial y}\Big|_c\right)\sin \alpha \ . \tag{A.9}$$

Now, let's look at a couple of particularily interesting cases. Suppose we wanted to select the particular line for which $\frac{d\phi}{ds}\Big|_c = 0$ equation (A.9) tells us that the angle (α_1) orienting this line is given by

$$\tan \alpha_1 = -\frac{\left(\frac{\partial\phi}{\partial x}\Big|_c\right)}{\left(\frac{\partial\phi}{\partial y}\Big|_c\right)} \qquad \text{("contour direction")} \tag{A.10}$$

Now, lets also find the particular direction which results in the maximum or minimum $\frac{d\phi}{ds}\Big|_c$ {the direction of "steepest ascent" (descent)}. The necessary condition for an extrema of $\frac{d\phi}{ds}\Big|_c$ requires

$$\frac{d}{d\alpha}\left(\frac{d\phi}{ds}\Big|_c\right) = 0 - \left(\frac{\partial\phi}{\partial x}\Big|_c\right)\sin \alpha + \left(\frac{\partial\phi}{\partial y}\Big|_c\right)\cos \alpha$$

from which the $\alpha = \alpha_2$ which orients the direction of steepest ascent (descent) is given by

$$\tan \alpha_2 = \frac{\left(\frac{\partial \phi}{\partial y}\Big|_c\right)}{\left(\frac{\partial \phi}{\partial x}\Big|_c\right)}$$ ("direction of steepest ascent (descent)" or "gradient direction") (A.11)

Notice that $(\tan \alpha_1) \cdot (\tan \alpha_2) = -1$ therefore α_1 and α_2 orient lines that are perpendicular ... the contour line is perpendicular to the gradient line ... see sketch of figure A2.

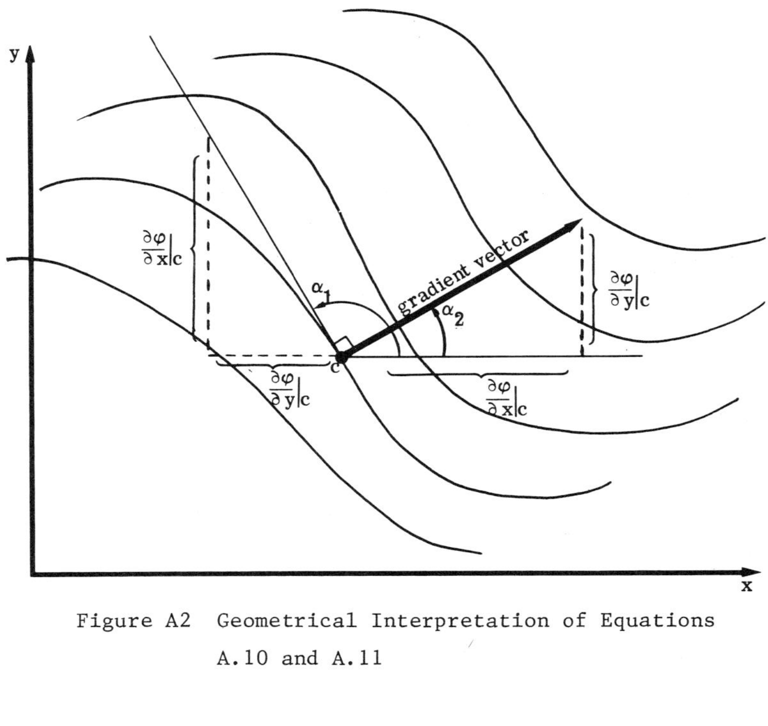

Figure A2 Geometrical Interpretation of Equations A.10 and A.11

These geometrical concepts are difficult to conceptualize rigorously in higher dimensional spaces (ever try to draw a 4 dimensional picture?!), but fortunately, the mathematics does generalize rigorously and in a straight-forward fashion.

A.4.2 Method of Gradients

One immediate conclusion of the foregoing is that (based only upon first derivative information), the most favorable direction to take a small step toward maximizing (or minimizing) the function ϕ is up (or down) the locally evaluated gradient of ϕ. The "method of gradients" (or the "method of steepest ascent" for maximizing ϕ or the "method of steepest descent" for minimizing ϕ) is a sequence of one dimensional searches along the lines established by successively evaluated local gradients of ϕ. Consider ϕ to be a function of n variables which are the elements of an n-vector X. Let the local evaluations be denoted by superscripts, e.g.

$$\phi^{(k)} = \phi(X^{(k)})$$

denotes $\phi(x_1,x_2,\ldots,x_n)$ evaluated at the kth set of x_i-values.

The kth one dimensional search determines a scalar $\alpha^{(k)}$ such that

$$X^{(k+1)} = X^{(k)} + \alpha^{(k)} \{\nabla_x\phi\}^{(k)} \tag{A.12}$$

results in

$$\phi^{(k+1)} = \phi(X^{(k+1)}) \tag{A.13}$$

being a local maximum (minimum), where

$$X \equiv \begin{Bmatrix} x_1 \\ \vdots \\ x_n \end{Bmatrix} \tag{A.14}$$

$$\nabla_x\phi \equiv \begin{Bmatrix} \dfrac{\partial\phi}{\partial x_1} \\ \vdots \\ \dfrac{\partial\phi}{\partial x_n} \end{Bmatrix} \tag{A.15}$$

It is important to develop a geometrical feel for the method of gradients to understand the circumstances under which it works best, to anticipate failures, and to decide upon remedial action when failure occurs. For these reasons, consider the sequence of applications of the gradient algorithm sketched in figures A3a, b, c, d.

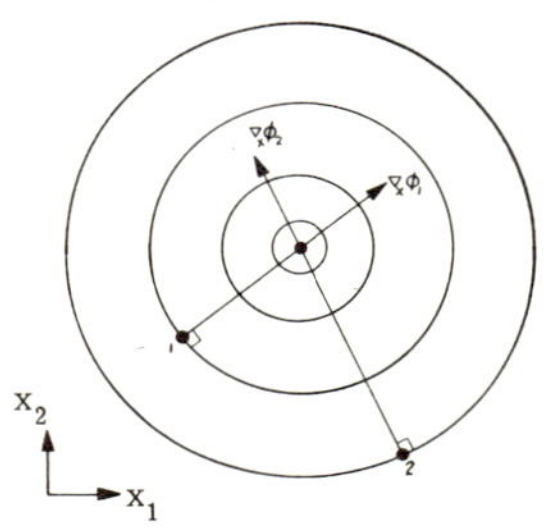

Fig. A3a. Circular Contours: Optimum Convergence

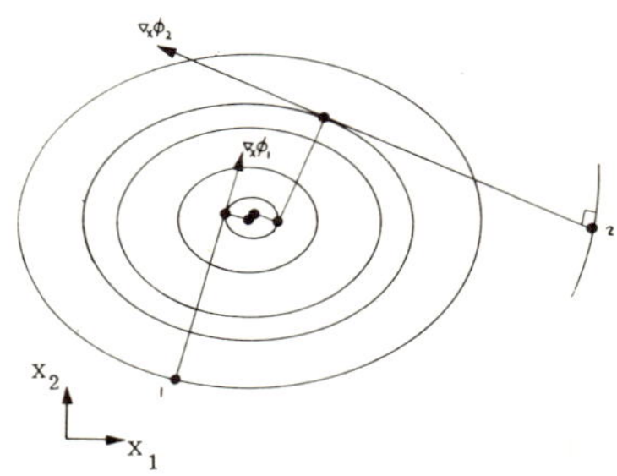

Fig. A3b. Low eccentricity contours: Good Convergence

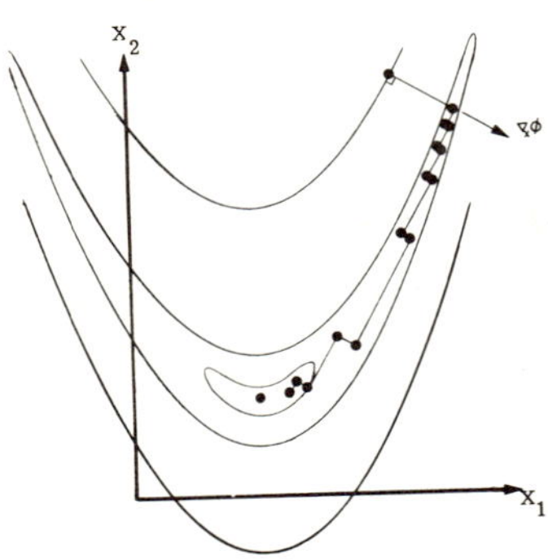

Fig. A3c. Nonlinear Trenches or Ridges: Poor Convergence

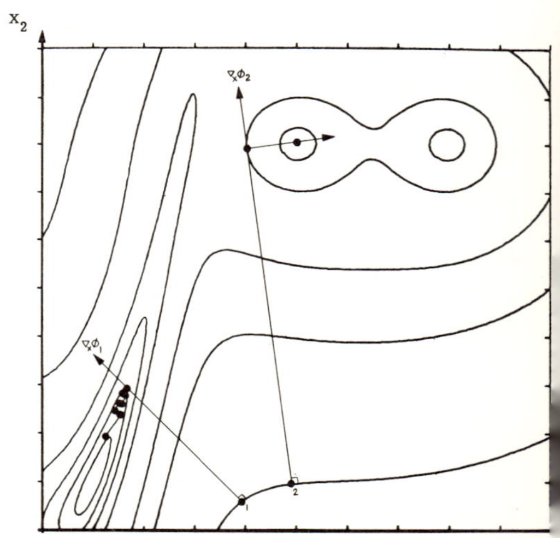

Fig. A3d. Multimodel Surfaces Local Extrema

Figure A3 The Convergence of the Method of Gradients Versus the Surface Regularity

It is clear that the convergence of the gradient algorithm is heavily dependent upon the circularity of the contours (In 3-space, the "contours" most desired are spherical surfaces; in n-space, the "contours" most desired are "hyperspheres").

Observe the orthogonality of successive gradients. The successive gradients will be exactly orthogonal only if the one-dimensional searches are done in such a fashion that one dimensional maxima or minima are perfectly located. Note, for the two dimensional case, that *only one gradient calculation may be necessary*, since all successive gradients are either parallel or perpendicular to the first. Why is the orthogonality condition insufficient to establish the gradient directions if the dimension of the space is greater than 2?

The gradient method often converges rapidly for the first few iterations (far from the solution), but is usually a very poor algorithm during the final iterations. For any function ϕ with non-spherical contours, the number of iterations to converge exactly is generally unbounded. Satisfactory convergence accuracy often requires an unacceptably large number of one-dimensional searches.

Exercise A1

Given the starting estimates

$$\begin{Bmatrix} x_1 \\ x_2 \end{Bmatrix}^{(1)} = \begin{Bmatrix} 1000 \\ 1000 \end{Bmatrix}$$

use the method of gradient to minimize

$$\phi(x_1,x_2) = \sum_{j=1}^{3} \left(a_j - \tan^{-1}\left(\frac{x_2}{x_1-b_j}\right) \right)^2 c_j$$

Given also the data

j	a_j	b_j	c_j
1	30.1	0	100
2	45.0	500	100
3	73.6	1000	25

Carry out four one dimensional searches using (A.12); use any convenient method you wish to find the optimum α-values, in each 1-d search.

See how rapidly your solution converges toward the optimum solution

$$\begin{Bmatrix} \hat{x}_1 \\ \hat{x}_2 \end{Bmatrix} = \begin{Bmatrix} 1204.783 \\ 701.775 \end{Bmatrix}$$

where

$$\phi(\hat{x}_1, \hat{x}_2) = 3.3892.$$

(see example problem 3.2.1)

Motivated by the poor terminal convergence of gradient algorithms, one "might consider the wisdom" of using higher derivatives in the local decision process which determines successive iterations.

A.4.3 Second Order (Gauss-Newton) Algorithm

The Gauss-Newton algorithm is probably the most powerful unconstrained optimization method. We will discuss a "curvature pitfall" which necessitates care in applying this algorithm, however.

A payoff function ϕ is evaluated at a local point $X^{(k)}$. It is desired to modify $X^{(k)}$ by $\Delta X^{(k)}$ according to

$$X^{(k+1)} = X^{(k)} + \Delta X^{(k)} \qquad (A.16)$$

in such a fashion that ϕ is increased or decreased. The behavior of ϕ near $X^{(k)}$ can be approximated by the second order Taylor's series

$$\phi(x) \cong \phi(X^{(k)}) + \{G^{(k)}\}\Delta X + \tfrac{1}{2}\Delta X^t H^{(k)} \Delta X \qquad (A.17)$$

where

$$G^{(k)} \equiv \nabla_x \phi^{(k)} \equiv \begin{Bmatrix} \dfrac{\partial\phi}{\partial x_1} \\ \vdots \\ \dfrac{\partial\phi}{\partial x_n} \end{Bmatrix}^{(k)} = \text{"gradient of } \phi\text{"} \qquad (A.18)$$

$$H^{(k)} \equiv \nabla_X^2 \phi^{(k)} \equiv \begin{bmatrix} \frac{\partial^2\phi}{\partial x_1^2} & \cdots & \frac{\partial^2\phi}{\partial x_1 \partial x_n} \\ \vdots & \ddots & \vdots \\ \frac{\partial^2\phi}{\partial x_n \partial x_1} & \cdots & \frac{\partial^2\phi}{\partial x_n^2} \end{bmatrix}^{(k)} = \text{"Hessian of } \phi\text{"} \tag{A.19}$$

The local strategy is to determine the particular correction vector $\Delta X^{(k)}$ which maximizes (minimizes) the second order prediction (A.17) of ϕ. Investigating (A.17) for an extreme leads to the following

<u>necessary conditions</u>

$$\nabla_{\nabla X}\phi = \{G^{(k)}\} + \{H^{(k)}\}\Delta X = 0 \tag{A.20}$$

<u>sufficient conditions</u>

$$\nabla^2_{\Delta X}\phi = \{H^{(k)}\} \begin{cases} \text{must be pos. def. for min.} \\ \text{must be neg. def. for max.} \\ \text{must be indefinite for saddle} \end{cases} \tag{A.21}$$

From the necessary conditions, the local corrections are then

$$\Delta X^{(k)} = - \{H^{(k)}\}^{-1} \{G^{(k)}\} \tag{A.22}$$

Equation (A.22) allows us to immediately set down {using (A.16)} the Gauss-Newton 2nd order optimization algorithm as simply

$$X^{(k+1)} = X^{(k)} - \{H^{(k)}\}^{-1}\{G^{(k)}\} \tag{A.23}$$

As an exercise, apply the above iteration process (three times) to exercise A1 and compare with the corresponding gradient algorithm iterations.

It is important that algorithm (2) converges in exactly one iteration for a quadratic payoff function; regardless of the starting estimates used. In many (probably most) solvable unconstrained optimization problems, the second order approximation underlying (A.23) becomes valid during the final iterations; the terminal convergence of (A.23) is usually exceptionally rapid.

There is a pitfall! If the sufficient condition (A.21) is not satisfied, then the correction will be in the wrong direction. It is difficult to attempt maximizing a function by solving for local minima. This pitfall can be circumvented by using a gradient algorithm until the neighborhood of the solution is reached, then testing the sufficient conditions (A.21); and employing the second order algorithm if it is satisfied.

The overwhelmingly most significant drawback of the above second order method is the necessity of calculating the matrix of second derivatives. For complicated payoff function models, it is usually an expensive consideration to simply determine the n elements of the gradient vector. One is thus motivated to ask the question: "Is it possible to approximate quadratic convergence without the expense of calculating the second partial derivatives?" The answer turns out to be yes! Observe that some "second order information" is contained in a sequence of local function and gradient calculations. Two such techniques have been developed since the mid-1960's and are in common use {the Fletcher-Reeves (1964) and Fletcher-Powell (1963) algorithms}.

These algorithms are not developed here due to space limitations, the interested reader should see Fiacco and McCormick (1968) for theoretical development and numerical examples of these important algorithms.

It is also significant to note that when the payoff function is the sum of squares of a set of functions whose first derivatives are available, that second order convergence can be approximated by linearizing the functions *before squaring*. The result is a local quadratic approximation of ϕ; this local approximation can be minimized rigorously. The classical example use of this device is the *Gaussian least square differential correction algorithm*. This algorithm is developed in §1.5 of the text and is applied to numerous examples in this text.

References for Appendix A

A.1 Fletcher, R., and M.J.D. Powell, "A Rapidly Convergent Descent Method for Minimization", Computer Journal 6, pp 163-168 (1963).

A.2 Fletcher, R., and C.M. Reeves, "Function Minimization by Conjugate Gradients", Computer Journal 7, 149-154 (1964).

A.3 Fiacco, A.V., and G.P. McCormick, Nonlinear Programming: Sequential Unconstrained Minimization Techniques, Wiley, New York, Ch 8 (1968).

A.4 Bryson, A., and Ho, Y., Applied Optimal Control, Blaisdell, London, Ch 10 (1969).

APPENDIX B

Basic Probability Concepts

B.1 Functions of a Single Discrete-Valued Random Variable

To appeal to the intuitive feel that we have for random variables and elementary probability concepts, attention is first directed to a simple experiment. Consider a single throw of a "true" die; the *probability* of the occurrence of each of the *events* 1, 2, 3, 4, 5, or 6 is exactly the same on a given throw. For a "loaded" die, the probability of certain of the events would be greater than others. If a given discrete-valued experiment is conducted N times and N_j is the number of times that the jth event x(j) occurred; it is then intuitively reasonable to define the probability of the occurrence of x(j) as

$$p(x(j)) \triangleq \lim_{N\to\infty} \frac{N_j}{N} \ . \tag{B.1}$$

A compound event can be defined as the occurrence of "either X(j) or X(k)"; the probability of the compound event is defined as

$$p(x(j)) \ \mathit{or} \ x(k)) = \lim_{N\to\infty} \frac{N_j+N_k}{N} \ . \tag{B.2}$$

From these definitions, it is clear that a *probability mass function* p(x(j)) has the following properties.

$$\sum_j p(x(j)) = 1 \tag{B.3}$$

$$0 < p(x(j)) < 1 \tag{B.4}$$

$$p(x(j)) \ \mathit{or} \ x(k)) = p(x(j)) + p(x(k)) \tag{B.5}$$

A *discrete-valued random variable*, x, is defined as a function having a finite number of possible values x(j); with the associated probability of x(j) occurring being denoted p(x(j)). To compact notation, x(j) and p(x(j)) are hereafter called x and p(x), whenever this substitution does not cause ambiguity.

The random variable x is usually described in terms of its *moments*. The first two moments of x are

The *mean* ($\bar{x}$) of x:

$$\bar{x} \triangleq \sum_j x(j)p(x(j)) \tag{B.6}$$

The *variance* (σ_x^2) of x:

$$\sigma_x^2 \triangleq \sum_j (x(j) - \bar{x})^2 p(x(j)). \tag{B.7}$$

The quantity σ_x is often called the *standard deviation* of x.

If p(x) is considered to be a function defining the mass of several discrete masses located along a straight line, then $\bar{x}$ locates the center of mass and σ_x^2 is the moment of inertia of the system of masses about their centroid.

The *expected value* or "average value" of a function f(x) of a discrete random variable x is defined as

$$E\{f(x)\} \triangleq \sum_j f(x(j))p(x(j)) \tag{B.8}$$

Clearly from equations (B.6) and (B.7), the mean and variance are the expected values of x and $(x-\bar{x})^2$, respectively. Notice that the expected value operator is linear so that

$$E\{af(x) + bg(x)\} = aE\{f(x)\} + bE\{g(x)\}, \tag{B.9}$$

for a and b arbitrary deterministic scalars; f(x) and g(x) arbitrary functions of the random variable x.

B.2 Functions of Discrete-Valued Random Vectors

A random vector X is an n x 1 matrix whose elements x_i are scalar random variables as discussed in §B-1. If each scalar element x_i of X can take on a finite number (m_i) of discrete values $x_i(j_i)$, for $j_i = 1,2,\ldots,m_i$, then there are $m_1 m_2 \ldots m_n$ possible vectors. For a complete probabilistic characterization of X, its *joint probability function* $p(j_1,j_2,\ldots,j_n)$ must be specified, where $p(j_1,j_2,\ldots,j_n)$ is the probability that x_1 has its j_1^{th} value, x_2 has its j_2^{th} value, ..., x_n has its j_n^{th} value. The function $p(j_1,j_2,\ldots,j_n)$ is often written $p(x_1,x_2,\ldots,x_n)$ when no ambiguity results. On some occasions, one is interested in the *marginal probability mass function* given by

$$p(j_1) = \sum_{j_2=1}^{m_2} \sum_{j_3=1}^{m_3} \cdots \sum_{j_n=1}^{m_n} p(j_1,j_2,\ldots,j_n). \tag{B.10}$$

Note that $p(j_1)$ is the probability of a compound event; that x_1 takes on its j_1^{th} value while $x_2,x_3,\ldots,x_n$ take on any arbitrary possible values. Thus, a scalar random variable may represent an elementary or compound event, depending upon the dimension of the underlying space of events.

The marginal probability functions (B.10) are sufficient to fully probabilistically characterize the components of X, but to fully characterize X, it is necessary to specify $p(x_1,x_2,\ldots,x_n)$. As in the scalar case, it is customary to describe $p(x_1,x_2,\ldots,x_n)$ and X in terms of the moments of X. The first two moments are:

The *mean* $(\bar{X})$ of X:

$$\bar{X} \triangleq E\{X\} \equiv \begin{pmatrix} E(x_1) \\ \vdots \\ E(x_n) \end{pmatrix} \equiv \sum_{j_1=1}^{m_1} \cdots \sum_{j_n=1}^{m_n} \begin{pmatrix} x_1(j_1) \\ \vdots \\ x_n(j_n) \end{pmatrix} p(j_1,j_2,\ldots,j_n) \tag{B.11}$$

The *covariance* (Λ_{xx}) of X:

$$\Lambda_{xx} \triangleq E\{(X - \bar{X})(X - \bar{X})^t\} \tag{B.12}$$

$$\Lambda_{xx} \equiv E\begin{bmatrix} (x_1-\bar{x}_1)^2 & (x_1-\bar{x}_1)(x_2-\bar{x}_2) & \cdots & (x_1-\bar{x}_1)(x_n-\bar{x}_n) \\ (x_2-\bar{x}_2)(x_1-\bar{x}_1) & (x_2-\bar{x}_2)^2 & \cdots & (x_2-\bar{x}_2)(x_n-\bar{x}_n) \\ \vdots & \vdots & \ddots & \vdots \\ (x_n-\bar{x}_n)(x_1-\bar{x}_1) & (x_n-\bar{x}_n)(x_2-\bar{x}_2) & \cdots & (x_n-\bar{x}_n)^2 \end{bmatrix}$$

$$\equiv \sum_{j_1=1}^{m_1} \cdots \sum_{j_n=1}^{m_n} \begin{bmatrix} (x_1(j_1) - \bar{x}_1)^2 & \cdots & (x_1(j_1)-x_1)(x_n(j_n)-\bar{x}_n) \\ \vdots & \ddots & \vdots \\ (x_n(j_n)-\bar{x}_n)(x_1(j_1)-\bar{x}_1) & \cdots & (x_n(j_n)-\bar{x}_n)^2 \end{bmatrix} \cdot p(j_1, \cdots j_n)$$

where the expectation operator E{ } when "operating" upon a matrix, operates upon each individual element as indicated. Notice that the covariance matrix $\Lambda_{xx} = \{M_{ij}\}$ is symmetric. We adopt the following notations:

$$\mu_{ii} \equiv \sigma_{xi}^2 = E\{(X_i-\bar{X}_i)^2\} = \text{variance of } X_i \tag{B.13}$$

$$\mu_{ij} = E\{(X_i-\bar{X}_i)(X_j-\bar{X}_j)\} = \text{covariance of } X_i \text{ and } X_j \tag{B.14}$$

$$\rho_{ij} = \frac{\mu_{ij}}{\sigma_{x_i}\sigma_{x_j}} = \text{correlation of } X_i \text{ and } X_j. \tag{B.15}$$

Example:

Consider a vector with two components

$$X = \begin{Bmatrix} x_1 \\ x_2 \end{Bmatrix}$$

Suppose that the first component has two possible values

$x_1(1) = 0$

$x_1(2) = 10$

Suppose that the second component has three possible values

$x_2(1) = -10$

$x_2(2) = 0$

$x_2(3) = 10$

Suppose, further, that the six possible events have the following probabilities:

$$p(0,-10) = .1, \quad p(0,0) = .4, \quad p(0,10) = .1$$

$$p(10,-10) = .1, \quad p(10,0) = .1, \quad p(10,10) = .2$$

The expected value (mean) of X then follows from (B.11) as

$$\bar{X} = E\begin{Bmatrix} x_1 \\ x_2 \end{Bmatrix} = .1\begin{Bmatrix} 0 \\ -10 \end{Bmatrix} + .4\begin{Bmatrix} 0 \\ 0 \end{Bmatrix} + .1\begin{Bmatrix} 0 \\ 10 \end{Bmatrix} + .1\begin{Bmatrix} 10 \\ -10 \end{Bmatrix} + .1\begin{Bmatrix} 10 \\ 0 \end{Bmatrix} + .2\begin{Bmatrix} 10 \\ 10 \end{Bmatrix}$$

therefore

$$\bar{X} = \begin{Bmatrix} 4 \\ 1 \end{Bmatrix}$$

Similarily, the covariance matrix follows from (B.12) as

$$\Lambda_{xx} = E \begin{bmatrix} (x_1-\bar{x}_1)^2 & (x_1-\bar{x}_1)(x_2-\bar{x}_2) \\ (x_2-\bar{x}_2)(x_1-\bar{x}_1) & (x_2-\bar{x}_2)^2 \end{bmatrix}$$

$$= .1\begin{Bmatrix} -4 \\ -11 \end{Bmatrix}\{-4 \ -11\} + .4\begin{Bmatrix} -4 \\ -1 \end{Bmatrix}\{-4 \ -1\} + .1\begin{Bmatrix} -4 \\ 9 \end{Bmatrix}\{-4 \ 9\}$$

$$+ .1\begin{Bmatrix} 6 \\ -11 \end{Bmatrix}\{6 \ -11\} + .1\begin{Bmatrix} 6 \\ -1 \end{Bmatrix}\{6 \ -1\} + .2\begin{Bmatrix} 6 \\ 9 \end{Bmatrix}\{6 \ 9\}$$

which reduces to

$$\Lambda_{xx} = \begin{bmatrix} 24 & 6 \\ 6 & 49 \end{bmatrix} \quad .$$

It may be verified from the results of Appendix A that this covariance matrix is positive definite.

To investigate the definitness of the definitness of Λ_{xx} in general, let

$$\bar{X} = E(X)$$

$z = C^t(X-\bar{X})$, C = an n x 1 matrix of arbitrary constrants

Investigating the moments of z, we find that

$$\bar{z} = E(z) = E\{C^t(X-\bar{X})\} = C^t(\bar{X}-\bar{X}) = 0$$

$$\begin{aligned}\sigma_z^2 = E\{(z-\bar{z})^2\} &= E\{(z)^2\} = E(zz^t) = E\{C^t(X-\bar{X})(X-\bar{X})^tC\} \\ &= C^tE\{(X-\bar{X})(X-\bar{X})^t\}C \\ &= C^t\Lambda_{xx}C\end{aligned}$$

Therefore

$$\sigma_z^2 = C^t\Lambda_{xx}C.$$

Since $\sigma_z^2 \geq 0$ and since C is an arbitrary vector, then Λ_{xx} is always *at least* positive semi-definite. For diagonal Λ_{xx}, the positive semi-definitess of Λ_{xx} agrees well with our intuitive interpretation of σ_{xi}^2; since $\sigma_{xi}^2 < 0$ implies "better than perfect knowledge" or "less than zero uncertainty" in x_i, which is impossible!

We have seen that, in general, it is not sufficient to know the probability functions of the components of a random vector X to fully characterize X. This fact manifests itself, for example, in the covariance matrix having non-zero off-diagonal terms. If $\mu_{ij} \neq 0$ for $i \neq j$, then x_i and x_j are said to be *correlated*. If

$$p(x_1,x_2,\ldots,x_n) = p(x_1)p(x_2)\ldots p(x_n) \tag{B.16}$$

for all possible values of $\{x_1,x_2,\ldots,x_n\}$, then the random variables $\{x_1,x_2,\ldots,x_n\}$ are said to be *independent*. Note that while pairwise independence is sufficient to insure zero correlation of $\{x_1,\ldots,x_n\}$, it is not sufficient to insure independence of $\{x_1,x_2,\ldots,x_n\}$.*

---

*See Feller, (1957), page 116.

B.3 Functions of Continuous Random Vectors

For our purposes, the discrete random variable concepts of §B.1 and B.2 can be extended in a natural manner.* By letting $N \to \infty$ with the probability mass function $p(x_1(j_1),\ldots, x_n(j_n))$ being replaced by a *probability density function* $p(x_1,\ldots,x_n)$; then

$$p(x_1,x_2,\ldots,x_n)dx_1,dx_2\ldots dx_n \tag{B.17}$$

is the probability that the components of X lie within the differential volume $dx_1,dx_2\ldots dx_n$ centered at $x_1,x_2,\ldots,x_n$. Since all possible X-vectors are located in the infinite sphere, it follows that

$$\int_{-\infty}^{\infty}\cdots\int_{-\infty}^{\infty} p(x_1,\cdots,x_n)dx_1\cdots dx_n = 1. \tag{B.18}$$

The expected value of an arbitrary function $g(x_1,\cdots,x_n)$ is defined in terms of the density function as

$$E\{g(x_1,\cdots,x_n)\} = \int_{-\infty}^{\infty}\cdots\int_{-\infty}^{\infty} g(x_1,\cdots,x_n)p(x_1,\cdots,x_n)dx_1\cdots dx_n \tag{B.19}$$

Thus the summation signs of the discrete results of §B.2 are replaced by integral signs to obtain the corresponding continuous results.

B.4 Propagation of Moments and Density Functions Through Linear and Nonlinear Models

B.4.1 Linear Scalar Models

Attention is first directed to the linear scalar equation

$$y = ax + b \tag{B.20}$$

where a and b are arbitrary deterministic scalars and x is a random variable whose first two moments

*There are various theoretical details that must be focused in a rigorous extension of the discrete results to the continuous results, see Parzen (1960), for example.

$$\bar{x} = E(x) \tag{B.21}$$

and

$$\sigma_x^2 = E\{(x-\bar{x})^2\} \tag{B.22}$$

are assumed known. It is desired to determine the resulting first two moments of y. The mean follows directly from the definition

$$\bar{y} = E(y)$$

upon substituting (B.20) as

$$\bar{y} = E(ax+b) ,$$

or

$$\bar{y} = a\bar{x} + b. \tag{B.23}$$

The variance of y can now be obtained as

$$\sigma_y^2 = E\{(y-\bar{y})^2\} ,$$

where, substituting from (B.20) and (B.23), we find

$$\sigma_y^2 = E\{a^2(x-\bar{x})^2\} \tag{B.24}$$

or

$$\sigma_y^2 = a^2\sigma_x^2 . \tag{B.25}$$

B.4.2 Linear Matrix Models

Attention is now directed to the linear matrix equation

$$Y = AX + B \tag{B.26}$$

where A and B are arbitrary constant matrices with deterministic elements, and X is a random vector whose first two moments

$$\bar{X} = E(X) \tag{B.27}$$

and

$$\Lambda_{xx} = E\{(X-\bar{X})(X-\bar{X})^t\} \tag{B.28}$$

are assumed known.

It is desired to determine the first and second moments of Y. The mean follows as

$$\bar{Y} = E(Y) = E(AX+B) = AE(X) + B$$

or

$$\bar{Y} = A\bar{X} + B \ . \tag{B.29}$$

The covariance matrix is then obtained from the definition

$$\Lambda_{yy} \equiv E\{(Y-\bar{Y})(Y-\bar{Y})^t\} \tag{B.30}$$

By substituting from (B.26) and (B.29) as

$$\Lambda_{yy} = E\{A(X-\bar{X})(X-\bar{X})^t A^t\} = AE\{(X-\bar{X})(X-\bar{X})^t\}A^t$$

or

$$\Lambda_{yy} = A\Lambda_{xx}A^t \tag{B.31}$$

which is a commonly used result for "mapping" covariance matrices through linear systems.

B.4.3 Transformation of Density Functions Through Non-Linear Models

If X is a random vector whose density function p(X) is known, and if Y = F(X) is an arbitrary (generally non-linear) 1-to-1 transformation, then it can be shown* that the density function of Y is given by

$$p(Y) = \left(\frac{1}{\left| \frac{\partial f_i}{\partial X_j} \right|} \right) p(X) \tag{B.32}$$

where X on the right hand side of (B.32) is

$$X = F^{-1}(Y) \tag{B.33}$$

---

*Bryson and Ho, Applied Optimal Control, p.301.

Example B1

We will now employ the above results using the linear scalar model

$$y = ax \tag{B.34}$$

and the assumed Gaussian density function for x

$$p(x) = \frac{1}{\sigma_x \sqrt{2\pi}} \exp\left(\frac{-x^2}{2\sigma_x^2}\right) , \tag{B.35}$$

then, applying (B.32) and (B.35), we find

$$p(y) = \frac{1}{a\sigma_x\sqrt{2\pi}} \exp\left(\frac{-y^2}{2a^2\sigma_x^2}\right) \tag{B.36}$$

Note further that

$$\bar{y} = E(y)$$

or

$$\bar{y} \equiv \int_{-\infty}^{\infty} yp(y)dy , \tag{B.37}$$

which, upon substituting from (B.36), we find

$$\bar{y} = \frac{1}{a\sigma_x\sqrt{2\pi}} \int_{-\infty}^{\infty} y \exp\left(\frac{-y^2}{2a^2\sigma_x^2}\right) dy . \tag{B.38}$$

Integrating (B.38) by parts, we find

$$\bar{y} = 0 \equiv \bar{x} . \tag{B.39}$$

Similarily, we find from the definition of variance that

$$\sigma_x^2 = E\{(y-\bar{y})^2\}$$

$$= E\ (y^2)$$

$$= \int_{-\infty}^{\infty} y^2p(y)dy$$

$$= \frac{1}{a\sigma_x\sqrt{2\pi}} \int_{-\infty}^{\infty} y^2 \exp\left(\frac{-y^2}{2a^2\sigma_x^2}\right) dy$$

which integrates to

$$\sigma_y^2 = a^2\sigma_x^2 \tag{B.40}$$

which confirms the previous result (B.25) for linear scalar systems.

Example B2

Assume the quadratic model

$$y = ax^2 \tag{B.41}$$

(note for each value of y there are 2 x-values), and assume x has the Gaussian density function

$$p(x) = \frac{1}{\sigma_x\sqrt{2\pi}} \exp\left(\frac{-x^2}{2\sigma_x^2}\right) . \tag{B.42}$$

It then follows from (B.32) and (B.33) that

$$p(y) = \frac{1}{2\sigma_x\sqrt{2\pi ay}} \exp\left(-\frac{y}{2a\sigma_x^2}\right) , \; y > 0 \tag{B.43}$$

and

$$p(y) = 0, \quad y < 0. \tag{B.44}$$

It also follows that

$$\bar{y} = E(y) = a\sigma_x^2 \tag{B.45}$$

and

$$\sigma_y^2 = E\{(y-a\sigma_x)^2\} = 2a^2\sigma_x^4 . \tag{B.46}$$

B.5 *Gaussian Density Function for a Random Vector*

The most commonly used density function, therefore most important (for our purposes) is the Gaussian or normal distribution,

$$p(X) = \frac{1}{(2\pi)^{n/2}|\Lambda_{xx}|^{\frac{1}{2}}} \exp\left[-\tfrac{1}{2}(X-\bar{X})^t\Lambda_{xx}^{-1}(X-\bar{X})\right] \tag{B.47}$$

It can be shown*

$$\int_{-\infty}^{\infty} \cdots \int_{-\infty}^{\infty} p(X)dx_1 \cdots dx_n = 1 \qquad \text{(B.48)}$$

$$\bar{X} = E(X) = \int_{-\infty}^{\infty} \cdots \int_{-\infty}^{\infty} Xp(X)dx_1 \cdots dx_n \qquad \text{(B.49)}$$

and

$$\Lambda_{xx} = E(X-\bar{X})(X-\bar{X})^t = \int_{-\infty}^{\infty} \cdots \int_{-\infty}^{\infty} (X-\bar{X})(X-\bar{X})^t p(X)dx_1 \cdots dx_n \qquad \text{(B.50)}$$

One is often interested in the probability that X lies inside the quadratic hypersurface

$$(X-\bar{X})^t \Lambda_{xx}^{-1}(X-\bar{X}) < R^2 \qquad \text{(B.51)}$$

where R is a constant. By an appropriate orthogonal transformation,**

$$Y = AX \qquad \text{(B.52)}$$

$$S \equiv \begin{pmatrix} \sigma_1^2 & \cdots & 0 \\ \vdots & \ddots & \vdots \\ 0 & \cdots & \sigma_n^2 \end{pmatrix} = A\Lambda_{xx}^{-1} A^t \qquad \text{(B.53)}$$

it is always possible to transform coordinates to a principal system in which (B.51) reduces to

$$\frac{y_1^2}{\sigma_1^2} + \frac{y_2^2}{\sigma_2^2} + \cdots + \frac{y_n^2}{\sigma_n^2} < R^2 \qquad \text{(B.54)}$$

By one further change of variables

$$z_i = \frac{y_i}{\sigma_i}, \quad i = 1,2,\cdots,n, \qquad \text{(B.55)}$$

then (B.54) becomes the equation of an n-dimensional sphere

$$z_1^2 + z_2^2 + \cdots + z_n^2 < R^2 \qquad \text{(B.56)}$$

---

* See Raiffa and Schlaifer (1961), pp. 246-251.

**The rows or columns of A are taken as the unit eigenvectors of Λ_{xx}^{-1}.

The probability of finding z inside this hypersphere is obtained by integrating the Gaussian density function over the volume of the sphere (B.56) as

$$P(\Sigma z_i^2 \leqq R^2) = \int_V p(z)dV \tag{B.56}$$

Using the element of volume $dz_1, dz_2 \cdots dz_n$, equation (B.56) can be written (noting $\Lambda_{zz} = I$) as

$$P(\sum_{i=1}^{n} z_i^2 \leqq R^2) = \int \cdots \int_V \frac{1}{(2\pi)^{n/2}} \exp \tfrac{1}{2}(z_1^2 + \quad + z_n^2) dz_1 \cdots dz_n, \tag{B.57}$$

or, using the substitution

$$r^2 = z_1^2 + z_2^2 + \cdots + z_n^2$$

and a n-dimensional spherically symmetric volume element f(r)dr, then (B.57) can be written as

$$P(r^2 \leqq R^2) = \frac{1}{(2\pi)^{n/2}} \int_0^R \exp(-\tfrac{1}{2}r^2) f(r)dr \tag{B.58}$$

For n = 1,2,3, (B.58) is explicitely:

$n = 1$, $f(r)dr = 2dr$

$$P(r \leqq R) = \sqrt{\frac{2}{\pi}} \int_0^R \exp(-\tfrac{1}{2}r^2)dr = \text{erf}\left(\frac{R}{\sqrt{2}}\right) \tag{B.59}$$

$n = 2$, $f(r)dr = 2\pi r dr$

$$P(r \leqq R) = \int_0^R \exp(-\tfrac{1}{2}r^2) r dr = 1 - \exp\left(\frac{-R^2}{2}\right) \tag{B.60}$$

$n = 3$, $f(r)dr = 4\pi r^2 dr$

$$P(r \leqq R) = \sqrt{\frac{2}{\pi}} \int_0^R \exp(-\tfrac{1}{2}r^2) r^2 dr = \text{erf}\left(\frac{R}{\sqrt{2}}\right) - \sqrt{\frac{2}{\pi}} R\exp\left(\frac{-R^2}{2}\right) \tag{B.61}$$

The numerical value of $P(r < R)$ is often of particular interest in error analysis. The following table displays the "curse of dimensionality" for the probability of r being within 1,2, and 3 "sigma ellipsoids" for 1, 2, and 3 dimensional spaces.

n \ R	1	2	3
1	0.683	0.955	0.997
2	0.394	0.865	0.989
3	0.200	0.739	0.971

Probability that $r \leq R$

Example B3

Consider a normally distributed two-dimensional vector with zero mean and covariance matrix

$$\Lambda_{xx} = \begin{bmatrix} 4 & 1 \\ 1 & 1 \end{bmatrix}$$

The eigenvalues are the roots of the characteristic equation

$$\begin{bmatrix} 4 - \sigma^2 & 1 \\ 1 & 1 - \sigma^2 \end{bmatrix} = \sigma^4 - 5\sigma^2 + 3 = 0$$

which are

$\sigma_1^2 = 4.30$

$\sigma_2^2 = 0.70$

The eigenvectors follow from

$$\begin{bmatrix} 4 & 1 \\ 1 & 1 \end{bmatrix} e_1 = \sigma_1^2 e_1$$

and

$$\begin{bmatrix} 4 & 1 \\ 1 & 1 \end{bmatrix} e_2 = \sigma_2^2 e_2$$

as

$$e_1 = \begin{Bmatrix} 1 \\ .30 \end{Bmatrix} \text{ and } e_2 = \begin{Bmatrix} 1 \\ -3.30 \end{Bmatrix}$$

The ellipses of constant likelihood are then

$$(x_1 x_2) \begin{bmatrix} 4 & 1 \\ 1 & 1 \end{bmatrix}^{-1} \begin{Bmatrix} x_1 \\ x_2 \end{Bmatrix} = R^2$$

or

$$x_1^2 - 2x_1x_2 + 4x_2^2 = 3R^2$$

The ellipses (contours of constant propability) are shown below for R = 1, 2, and 3. From the above table (n = 2), the probability of finding X inside the R = 1, 2, and 3σ ellipses are 0.394, 0.865, and 0.989, respectively.

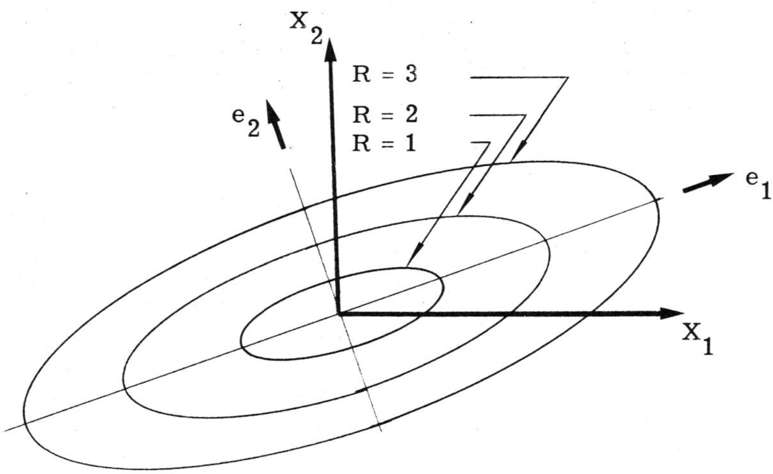

Figure B. 1 Constant Liklihood Ellipses for $E(X) = 0$, $E(XX^t) = \begin{pmatrix} 4 & 1 \\ 1 & 1 \end{pmatrix}$

B.6 Concluding Comments

This appendix serves as an overview of the probability concepts which are most important in the present text's approach to estimation theory. These developments were patterned after the excellent survey provided by Bryson and Ho (1969); the interested student is strongly encouraged to study probability theory formally from a conventional text such as Parzen (1960).

References for Appendix B

B.1 Feller, W., An Introduction to Probability Theory and Its Applications, Wiley, New York, p.116 (1967).

B.2 Parzen, E., Modern Probability Theory and Its Applications, Wiley, New York, ch 1-5 (1960).

B.3 Bryson, A., and Ho, Y., Applied Optimal Control, Blaisdell, London, ch 10 (1969).

B.4 Raiffa, H., and R. Schlaifer, Applied Statistical Decision Theory, Harvard Univ. Press, pp 246-251 (1961).

APPENDIX C

Linear Algebraic Equations

This appendix provides a reasonably comprehensive account of the most fundamental methods of linear algebra which impact the matrix calculations of estimation theory. However, most of the theorems are not proven here; those proofs given are *constructive* (i.e., suggest an algorithm). The account here is thus not satisfactorily self-contained, but references are provided where rigorous theorem proofs may be found. The motivation for this appendix is to clearly set down the most important algorithms and to point out potential numerical pitfalls associated with these algorithms.

C.1 Linear Equations and Matrix Inverses

The system of m linear equations

$$\begin{aligned} y_1 &= a_{11}x_1 + a_{12}x_2 + \cdots + a_{1n}x_n \\ y_2 &= a_{21}x_1 + a_{22}x_2 + \cdots + a_{2n}x_n \\ &\vdots \\ y_m &= a_{m1}x_1 + a_{m2}x_2 + \cdots + a_{mn}x_n \end{aligned} \tag{C.1}$$

can be written in matrix form as

$$Y = AX. \tag{C.2}$$

We consider conditions under which (C.2) has a solution for X given A and Y, and the closely related notion of a matrix inverse. The results of the current discussion are not constructive, (they do not imply computational algorithms), however, algorithms are developed through the remaining sections of this appendix.

The following terminology should be noted carefully: if $m > n$, the system (C.1) or (C.2) is said to be *overdetermined* (there are more equations than unknowns). Under typical circumstances we will find that exact solutions of (C.1) do not exist; therefore algorithms for *approximate* solutions for X are usually characterized by some measure of *how well* equations (C.1) are satisfied. If $m < n$, the system (C.1) or (C.2) is said to be *underdetermined* (there are fewer equations than unknowns). Under typical circumstances, an infinity of exact solutions for X exist; therefore solution algorithms have implicit some criterion for selecting a particular solution from the infinity of possible or feasible X solutions. If $m = n$ the system is said to be *determined*, under typical (but certainly not universal) circumstances, a unique exact solution for X exists.

The issues of whether or not solutions of (C.1) for X exist and whether or not these solutions are unique is treated in detail by Stewart (1973). Algorithms for the overdetermined case will be considered in subsequent discussions, for the moment consider the *determined* ($m = n$) case. The following theorems contains the fundamental truths for this case.

Theorem C.1 Let A be n x n matrix. The following statements are equivalent:

1. rank (A) = n
2. null (A) = 0
3. A has linearly independent columns
4. A has linearly independent rows
5. there is an *inverse matrix* A^{-1} of order n satisfying

$$A^{-1}A = AA^{-1} = I \quad . \tag{C.3}$$

Theorem C.2 Let A, B, and C be n x n matrices.

1. If $AB = I$, then $BA = I$, $B = A^{-1}$
2. If $AC = I$, then $B = C = A^{-1}$

Theorem C.3 A *nonsingular* matrix is a matrix whose inverse exists. If A is nonsingular then A^{-1} is nonsingular and

$$(A^{-1})^{-1} = A \; . \tag{C.4}$$

Likewise A^t is nonsingular and

$$(A^t)^{-1} \equiv (A^{-1})^t \tag{C.5}$$

Theorem C.4 Let A and B be n x n matrices, the matrix product AB is nonsingular if and only if A and B are nonsingular and

$$(AB)^{-1} = B^{-1} A^{-1} \tag{C.6}$$

These four theorems are proven in detail by Stewart (1973), we do not repeat the proofs here.

C.2 Elementary Row Operations

Many algorithms for matrix calculations center around *elementary row operations*. These operations, performed upon an arbitrary matrix A, are

1. multiply the ith row of A by a scalar α,
2. interchange rows i and j,
3. add α times row i to row j.

These operations can be accomplished by premultiplying A by suitable elementary matrices. The following matrices accomplish the three elementary row operations

1.
$$\begin{array}{c} \\ \\ i \\ \\ \end{array} \begin{pmatrix} 1 & & \overset{i}{\vdots} & & \\ & \ddots & \vdots & & \\ & & 1\vdots & & \\ & \cdots & \alpha & 1 & \\ & & & \ddots & \\ & & & & 1 \end{pmatrix} \tag{C.7}$$

2. $$\begin{array}{cc} & \begin{array}{ccccccc} & & i & & j & & \end{array} \\ \begin{array}{c} \\ \\ i \\ \\ j \\ \\ \\ \end{array} & \begin{pmatrix} 1 & & & & & & \\ & \ddots & & & & & \\ \cdots & 1 & 0 & \cdots & 1 & & \\ & & \vdots & 1 & \vdots & & \\ & & \vdots & \ddots & 1 & & \\ \cdots & & 1 & \cdots & 0 & & \\ & & & & & 1 & \\ & & & & & & \ddots \\ & & & & & & 1 \end{pmatrix} \end{array} \tag{C.8}$$

3. $$\begin{array}{cc} & \begin{array}{ccccc} & & i & j & \end{array} \\ \begin{array}{c} \\ \\ i \\ j \\ \\ \\ \end{array} & \begin{pmatrix} 1 & & & & \\ & \ddots & & & \\ \cdots & & 1 & \cdots 0 & \\ \cdots & & \alpha & \cdots 1 & \\ & & & & \ddots \\ & & & & & 1 \end{pmatrix} \end{array} \tag{C.9}$$

Notice that the matrix corresponding to an elementary row operation is the matrix obtained by performing the row operation on the identity matrix.

C.3 Matrix Reductions

Here we introduce a class of matrix operations which underlies many modern least squares algorithms and will be used in the subsequent developments of this appendix. A basic truth is captured by the following theorem:

Theorem C.5 Let A be an m x n matrix. There exists *nonsingular* matrices X and Y, of order n and m, respectively, such that

$$Y^tAX = \left(\begin{array}{c|c} C & 0 \\ \hline 0 & 0 \end{array}\right) \tag{C10}$$

where C is a nonsingular matrix of dimension $\ell \leq \min(m,n)$.

Stewart (1973) proves this theorem, but the proof is not constructive (i.e. does not establish an algorithm), so we omit the proof here. However, we establish algorithms for matrix reductions of the form (C.10); we will find that the *QR algorithm*, (most efficient and numerically stable algorithm for solution of the linear least squares problem) belongs to this class of reductions.

C.4 Classical Solutions and Reductions of Linear Systems

Many algorithms for solution of Y = AX have in common the idea of *factoring* A into the form

$$A = R_1\ R_2 \cdot\cdot\cdot\cdot\cdot R_k \tag{C.11}$$

where each matrix R_i is so simple that the equation Z = RP is easily solved for P. The original system Y = AX can then be solved by taking Z_o = Y, and for i = 1,2,···, k computing P_i as the solution of

$$Z_{i-1} = R_i P_i \tag{C.12}$$

Clearly P_k is the desired solution X.

In many modern algorithms, the factor matrices R_i are triangular. Hence we devote some attention to algorithms for solution of triangular systems of equations in §C.4.1. §C.4.2 and C.4.3 are devoted to developing algorithms for factoring A. §C.4.4 then applies these factorizations to solutions of linear systems.

When A is nonsingular, the methods of this appendix can be adapted to produce stable algorithms for calculating A^{-1}. It is important to stress that calculation of A^{-1} is *unnecessary* and *ill-advised* for applications requiring only a solution for X from equations C.2. However, there remain many applications in which the inverse matrix A^{-1} is itself a part of the *answer* desired (it turns out that the *covariance matrix* of the estimated parameters in batch and sequential least squares algorithms must be calculated from a matrix inverse, see §2.2 and 2.3). This latter point seems to have escaped the attention of recent authors of least squares algorithms.

C.4.1 Solution of Triangular Systems

An upper *triangular system* of linear equations has the form

$$\begin{aligned} t_{11}x_1 + t_{12}x_2 + t_{13}x_3 + \cdots + t_{1n}x_n &= y_1 \\ t_{22}x_2 + t_{23}x_3 + \cdots + t_{2n}x_n &= y_2 \\ t_{33}x_3 + \cdots + t_{3n}x_n &= y_3 \\ &\vdots \\ t_{nn}x_n &= y_n \end{aligned} \tag{C.13}$$

or

$$TX = Y \tag{C.14}$$

where T is the *upper triangular matrix*

$$T = \begin{bmatrix} t_{11} & t_{12} & t_{13} & \cdots & t_{1n} \\ 0 & t_{22} & t_{23} & \cdots & t_{2n} \\ 0 & 0 & t_{33} & \cdots & t_{3n} \\ \vdots & \vdots & & \ddots & \\ 0 & 0 & \cdots & & t_{nn} \end{bmatrix} \tag{C.15}$$

The following two theorems on the inverse of triangular systems are important in subsequent developments:

Theorem C.6 Let A and C be square matrices of order ℓ and m respectively, and let B be an ℓ x m matrix, with $\ell + m = n$ and

$$T = \begin{pmatrix} A & \vdots & B \\ \cdots & & \cdots \\ 0 & \vdots & C \end{pmatrix} \tag{C.16}$$

then the inverse of T exists and is

$$T^{-1} = \begin{pmatrix} A^{-1} & \vdots & -A^{-1}BC^{-1} \\ \cdots & & \cdots \\ 0 & \vdots & C^{-1} \end{pmatrix} \tag{C.17}$$

provided A and C are nonsingular.

Theorem C.7 Let T be upper triangular. Then T is nonsingular if and only if its diagonal elements are nonzero.

Theorems C.6 and C.7 are proven by Stewart (1973). It is apparent that the inverse of an upper triangular matrix is also upper triangular.

We now turn to solving the triangular system (C.13). From the final equation, it is apparent that

$$x_n = t_{nn}^{-1} y_n \tag{C.18}$$

In general, if we have calculated $x_n, x_{n-1}, \cdots, x_{i+1}$, we can solve the ith equation of (C.13) for x_i by simply dividing by x_i's coefficient as

$$x_i = t_{ii}^{-1} (y_i - \sum_{j=i+1}^{n} t_{ij} x_j) \tag{C.19}$$

This leads immediately to the *back substitution algorithm* for solution of triangular systems:

FOR i = n, n - 1, ···, 1

$$x_i = t_{ii}^{-1} (y_i - \sum_{j=i+1}^{n} t_{ij} x_j) \tag{C.20}$$

NEXT i

Clearly this algorithm can fail only if $t_{ii} \to 0$. But by theorem C.7, this can occur only if T is singular (or nearly singular). In practive, however, the accumulation of arithmetic errors is of concern *with any algorithm*. Experience indicates that algorithm (C.20) is well-behaved for *most* applications, and that double-precision acculumination of the parenthetic expression will *usually* eliminate loss of precision problems in the x_i, if a single precision result is satisfactory.

Note that algorithm (C.20) involves about $n^2/2$ multiplications.

Algorithm (C.20) can be modified to compute the inverse $S = T^{-1}$ of an upper triangular matrix T. Specifically, if S is partitioned by columns $S = (S_1 \vdots S_2 \vdots \cdots \vdots S_n)$, then the matrix equation

$$TS = I \tag{C.21}$$

is equivalent to solving the n sets of linear equations

$$TS_k = E_k \quad ; \; (k=1,2,\cdots,n) \tag{C.22}$$

where E_k is the kth column of the n x n identity matrix. However, solving the n linear systems (C.22) is inefficient, since it is known in advance that almost half of S's elements are zero and it does not make sense confirming then (except possibly for an expensive check on arithmetic errors). Let

$$S_k = \begin{Bmatrix} S_{1k} \\ S_{2k} \\ \vdots \\ S_{kk} \end{Bmatrix} \tag{C.23}$$

and E'_k be the kth column vector of the k x k identity matrix. Then it is easily verified that

$$T^{(k)} \; S'_k = E'_k \; ; \quad (k=1,2,\cdots,n) \tag{C.24}$$

where $T^{(k)}$ is the leading diagonal k x k sub-matrix of T. We now summarize an algorithm for calculating $S = T^{-1}$ and overwriting T by T^{-1}:

FOR $k = n, n-1, \cdots, 1$

$$t_{kk} \leftarrow S_{kk} = t_{kk}^{-1} \tag{C.25*}$$

$$t_{ik} \leftarrow S_{ik} = -t_{ii}^{-1} \sum_{j=i+1}^{k} t_{ij} \; S_{jk} \; ; \; (k = k-1, k-2, \cdots, 1)$$

NEXT k

---

*The symbol $x \leftarrow y$ means "overwrite" x by the current y-value. This notation is employed to indicate how storage may be conserved by overwriting quantities no longer needed.

The algorithm (C.25) involves about $n^3/6$ multiplications. Observe that the naive approach to solving (C.14) would be to first form

$$S = T^{-1}$$

using algorithm (C.25), then calculate the x-solution via the matrix multiplication X = SY.

The first step requires $n^3/6$ multiplications, in comparison to which the operations involved in the second step (X = SY) are negligible. However, if X is determined by the back-substitution algorithm (C.20), only $n^2/2$ multiplications are involved. Thus the multiplication count differs by the factor n/3 in favor of not calculating the matrix inverse! The question is not merely efficiency, it is also accuracy. As a rule of thumb, the fewer the operations, the less opportunity for aritheme tic errors and the accumulation thereof.

One can often avoid explicit calculation of matrix inverses in favor of solving a linear system by backsubstitution. As examples, consider the following:

Example C1

It is desired to compute

$$\alpha = X^t T^{-1} Y$$

to avoid calculating T^{-1}, one can first solve the linear system

$$TZ = Y$$

for Z, then calculate

$$\alpha = X^t Z.$$

Example C2

It is desired to calculate

$$C = T^{-1} B,$$

where C is n x n and B is n x m.

To avoid calculating T^{-1}, partition B and C by columns, then

$$\{C_1 \vdots C_2 \vdots \cdots \vdots C_m\} = \{T^{-1}B_1 \vdots T^{-1}B_2 \vdots \cdots \vdots T^{-1}B_m\} \quad ,$$

so that

$$C_k = T^{-1}B_k \ ; \ (k = 1,2,\cdots,m)$$

Thus the column vectors C_k can be determined by solving the m linear systems

$$T\ C_k = B_k \ ; \ (k = 1,2,\cdots,m)$$

by the back substitution algorithm (C.20).

C.4.2 Gaussian Elimination

Gaussian Elimination is a classical reduction procedure by which a square matrix A can be reduced to upper triangular form (thereby allowing solution by the methods of §C.4.1). More generally, Gaussian elimination reduces a m x n matrix to the *upper trapezodial form*

$$\begin{bmatrix} X & X & X & X & X & X \\ 0 & X & X & X & X & X \\ 0 & 0 & X & X & X & X \\ 0 & 0 & 0 & X & X & X \end{bmatrix}$$

where X's denote the non-zero entries.

Gaussian elimination involves premultiplications of the A matrix by a sequence of *elementary lower triangular* matrices. An elementary lower triangular matrix of order n and index k has the form

$$M = I_n - m\ E_k^t \tag{C.26}$$

where

$$E_i^t\ m = 0 \quad (i = 1,2,\cdots,k). \tag{C.27}$$

The conditions (C.27) dictate that the first k elements of column matrix m are zero or

$$m^t = (0\ 0 \cdots 0\ m_{k+1}\ m_{k+2} \cdots m_n) \tag{C.29}$$

In general, the elementary lower triangular matrix (C.26) has the form

$$M = \begin{bmatrix} 1 & 0 & \cdots & 0 & \cdots & 0 \\ 0 & 1 & \cdots & 0 & \cdots & 0 \\ \vdots & \vdots & & \vdots & & \vdots \\ 0 & 0 & \cdots & 1 & \cdots & 0 \\ 0 & 0 & \cdots & -m_{k+1} & \cdots & 0 \\ \vdots & \vdots & & \vdots & & \\ 0 & 0 & \cdots & -m_n & \cdots & 0 \end{bmatrix} \tag{C.30}$$

Thus M is an identity matrix except for the non-zero elements in the kth column below the diagonal. Elementary lower triangular matrices are easily inverted, as is indicated in the following theorem:

Theorem C.8 Let $M = I - m\ E_k^t$ be an elementary lower triangular matrix. Then

$$M^{-1} = I + m\ E_k^t\ . \tag{C.31}$$

The proof is trivial, let $X = I + M\ E_k^t$, then

$$\begin{aligned} M\ X &= (I - m\ E_k^t)\ (I + m\ E_k^t) \\ &= I - m\ E_k^t + m\ E_k^t + m\ E_k^t\ m\ E_k^t \\ &= I - m(E_k^t\ m)\ E_k^t \\ &= I, \text{ by virtue of equation (C.27).} \end{aligned}$$

The proof of the following theorem provides an algorithm for calculating an important class of elementary lower triangular matrices:

Theorem C.9 Let $E_k^t X = x_k \neq 0$, where $X^t = \{x_1 \cdots x_n\}$ and E_k is the kth column of I. There is a unique elementary lower triangular matrix M of index k such that

$$MX = \begin{Bmatrix} x_1 \\ \vdots \\ x_k \\ 0 \\ \vdots \\ 0 \end{Bmatrix} \tag{C.32}$$

Proof: We seek M in the form $M = I - m\,E_k^t$. Since M is to be of index k, we have

$$m_i = 0 \;;\; (i = 1,2,\cdots,k) \tag{C.33}$$

Observe

$$MX = (I - mE_k^t)\,X = X - (E_k^t\,X)\,m.$$

Since the last n - k components of MX are to be zero, we must have

$$x_i - x_k m_i = 0 \;;\; (i = k+1,k+2,\cdots,n) \tag{C.34}$$

and for $x_k = 0$, it is clear that

$$m_i = \frac{x_i}{x_k} \;;\; (i = k+1,k+2,\cdots,n). \tag{C.35}$$

Thus, if M exists, it is uniquely determined by equations (C.26), (C.33), and (C.35).

In the Gaussian elimination procedure, the A matrix is premultiplied by a sequence of lower triangular matrices, each chosen to introduce a column with zeros below the diagonal (this process is often called "annihilation"). The method makes repeated use of theorem C.9, which allows calculation of the appropriate elementary matrices.

Let $A_1 = A$ be an m x n matrix. If $a_{11}^{(1)} = a_{11}$ is non-zero, then by theorem C.9 there is a unique elementary lower triangular matrix M_1 of index 1 that annihilates the n-1 elements below the diagonal in the first column of A_1. Premultiplication of A_1 by M_1 produces a once-reduced matrix A_2 of the form

$$A_2 = M_1A_1 = \begin{bmatrix} a_{11}^{(1)} & a_{12}^{(1)} & a_{13}^{(1)} & \cdots & a_{1n}^{(1)} \\ 0 & a_{22}^{(2)} & a_{23}^{(2)} & \cdots & a_{2n}^{(2)} \\ 0 & a_{32}^{(2)} & a_{33}^{(2)} & \cdots & a_{3n}^{(2)} \\ \vdots & \vdots & \vdots & & \vdots \\ 0 & a_{m2}^{(2)} & a_{m3}^{(2)} & \cdots & a_{mn}^{(2)} \end{bmatrix} \tag{C.36}$$

At the kth stage of the reduction

$$A_k = M_{k-1}M_k \cdots M_1 A_1 , \tag{C.37}$$

A_k has the partitioned form

$$A_k = \left[\begin{array}{c:c} A_{11}^{(k)} & A_{12}^{(k)} \\ \hdashline 0 & A_{22}^{(k)} \end{array}\right] \tag{C.38}$$

Where $A_{11}^{(k)}$ is an upper triangular matrix of order k-1 and, for $i = 1,2,\cdots,k-1$, the ith row of A_k is the same as the ith row of A_i. The construction of M_k is a straight-forward application of theorem C.9.

Let the kth column of A_k be the vector

$$A_k^{(k)} = \begin{bmatrix} a_{1k}^{(1)} \\ a_{2k}^{(2)} \\ \vdots \\ a_{kk}^{(k)} \\ \vdots \\ a_{mk}^{(k)} \end{bmatrix} \tag{C.39}$$

and M_k is the elementary lower triangular matrix of index k such that

$$M_k\ A_k^{(k)} = \begin{bmatrix} a_{1k}^{(1)} \\ a_{2k}^{(2)} \\ \vdots \\ a_{2k}^{(k)} \\ 0 \\ \vdots \\ 0 \end{bmatrix} \tag{C.40}$$

Then by theorem C.9, the necessary elementary matrix is uniquely determined by

$$M_k = I - m^{(k)}\ E_k^t\ , \tag{C.41}$$

where

$$m^{(k)} = (0\ 0 \cdots 0\ m_{k+1}^{(k)} \cdots m_m^{(k)}) \tag{C.42}$$

and

$$m_j^{(k)} = \left(\frac{a_{jk}^{(k)}}{a_{kk}^{(k)}}\right) \quad ; \quad (j = k+1, k+2, \cdots, m) \tag{C.43}$$

provided $a_{kk}^{(k)} \neq 0$.

Considerable savings in storage and number of operations can be affected by taking advantage of the special structure of M_k. Instead of brute force calculation of A_{k+1} according to

$$A_{k+1} = M_k\ A_k$$

we make use of the fact that $M_k = I - m^{(k)}\ E_k^{(t)}$ to calculate A_{k+1} using

$$A_{k+1} = A_k - m^{(k)}\ E_k^t\ A_k \tag{C.44}$$

Since $E_k^t\ A_k$ is the kth row of A_k, then the ith row of A_{k+1} is calculated by subtracting $m_i^{(k)}$ times the kth row of A_k from the ith row of A_k. Since $m_1^{(k)} = m_2^{(k)} = \cdots = m_k^{(k)} = 0$, then only rows $k+1, k+2, \cdots m$ are modified. Thus the elements of the k+1 reduction can be calculated directly from

$$a_{ij}^{(k+1)} = a_{ij}^{(k)} - m_i^{(k)} a_{kj}^{(k)} ;$$

$$\text{FOR } (i = k+1, k+2, \cdots, m;\ j = k+1, k+2, \cdots, \cdots, n) \qquad (C.45)$$

with the understanding that $a_{ik}^{(k+1)} = 0$ for $i = k+1, k+2, \cdots, m$. Thus equations (C.43) and (C.45) are the explicit algorithm for calculating the k+1 step in the Gaussian elimination process. The process terminates whenever the rows or columns is exhausted. For $m > n$ this occurs after the nth reduction. Of $m < n$ this occurs on the (m-1)th reduction. Thus there are $r = \min(m-1, n)$ stages. The Gaussian elimination process may be implemented as the following algorithm:

For A an m x n matrix, $m > 1$, this algorithm overwrites A by the upper trapezoidal matrix resulting from the Gaussian elimination reductions.

$r = \min\{m-1, n\}$

FOR $k = 1, 2, \cdots, r$

$$\left.\begin{aligned} &a_{ik} \leftarrow m_i^{(k)} = \frac{a_{ik}}{a_{kk}} \ ; \ (i = k+1, \cdots, m) \\ &a_{ij} \leftarrow a_{ij} - m_i^{(k)} a_{kj} ; \quad i = k+1, \cdots, m \\ &\qquad\qquad\qquad\qquad\quad j = k+1, \cdots, n \end{aligned}\right. \qquad (C.46)$$

NEXT k

For large r, Gaussian elimination requires approximately $r^3/3$ multiplications. The divisors $a_{kk}^{(k)}$ are known as *pivot elements* in the Gaussian elimination process. Difficulty arises immediately if any of the pivot elements become zero. For obvious practical reasons, it would be desirable to understand how to prevent these pivot elements from tending toward zero. The following theorem provides one important truth:

Theorem C.10 The pivot elements $a_{ii}^{(i)}$; $(i = 1,2,\cdots,k)$ are non-zero if and only if the leading principal sub-matrices $A^{(i)}$ $(i = 1,2,\cdots,k)$ are all non-singular.

An induction proof of Theorem C.10 is provided by Stewart (1973). This theorem suggest that there may be profitable rearrangements of the original system to minimize the probability of encountering small pivot elements. This line of reasoning leads to an important computational device known as *pivoting*. At the kth stage of Gaussian elimination, a search can be carried out over the elements of the kth row and column, and through a suitable interchange of rows and columns, the element with the largest absolute value (of the sub-set of elements contained in the kth row and column) can be brought to the (k,k) pivot position. Thus the divisor in equation (C.43) can be maximized at each stage of the reduction. It is necessary to do some "bookkeeping" to remember the sequence of row and column interchanges so that the elements can be restored to their proper position (if desired) upon completion of the reduction process.

These ideas are carried to completion by Stewart (1973) who summarizes *Gaussian elimination with complete pivoting* as the following algorithm:

Let A be an m x n matrix with $m > 1$, let $r = \min\{m-1,n\}$. This algorithm overwrites A by the upper trapezoidal matrix A_{r+1} resulting from r Gaussian elimination stages with complete pivoting at each stage. The A_{r+1} matrix has it's rows and columns rearranged. The indices locating the original row and column positions are $\{\rho_1,\rho_2,\cdots,\rho_m\}$ and $\{\gamma_1,\gamma_2,\cdots,\gamma_n\}$, respectively.

FOR $k = 1,2,\cdots,r$

Find $\rho_k, \gamma_k \geq k$ such that

$$a_{\rho_k \gamma_k} = \max\{a_{ij}: i,j \geq k\}$$

If $a_{\rho_k \gamma_k} = 0$ set $r = k-1$ and halt the calculations

$$
\begin{aligned}
{}^{*}\ & a_{kj} \rightleftarrows a_{\rho_k, j} && (j = k, k+1,\cdots,n) \\
& a_{ik} \rightleftarrows a_{i,\gamma_k} && (i = 1,2,\cdots,m) \\
& a_{ik} \leftarrow m_{ik} = \frac{a_{ik}}{a_{kk}} && (i = k+1,\cdots,m) \\
& a_{ij} \leftarrow a_{ij} - m_{ik}a_{kj} && (i = k+1,\cdots,m;\ j = k+1,\cdots,n)
\end{aligned}
\tag{C.46}
$$

NEXT k

C.4.3 Triangular Decompositions

An important class of algorithms, in routine use since about 1960, involves factoring a square matrix A into a product LU of a lower and an upper triangular matrix. For obvious reasons, this factorization is called an LU decomposition of A. Actually, Gaussian elimination is a foremost example of LU decompositions. Observe from §C.4.2 that the result of the Gaussian elimination process, applied to a square matrix, was the upper triangular matrix

$$A_n = M_{n-1}M_{n-2}\cdots M_1 A \tag{C.47}$$

so that

$$A = M_1^{-1}M_2^{-1}\cdots M_{n-1}^{-1} A_n \tag{C.48}$$

and thus

$$A = LU \tag{C.49}$$

---

*The operation "$\rightleftarrows$" means "interchanges the value assigned to".

where

$$L = M_1^{-1} M_2^{-1} \cdots, M_{n-1}^{-1}$$

and

$$U = A_n \tag{C.51}$$

The $m_{ij} = m_i^{(j)}$ are the multipliers calculated via equations (C.43) during the Gaussian reduction process. Thus algorithm (C.46) can be regarded as a means for calculating and overwriting A by its LU decomposition with L being stored in the lower part of the array and $U = A_n$ in the upper part.

In general, the A = LU factorization of a matrix A is not unique. This can be seen by observing that for D an arbitrary nonsingular diagonal matrix that setting

$$L' = LD$$
$$U' = D^{-1}U$$

yield new upper and lower triangular matrices which satisfy the factorization requirement:

$$L'U' = LDD^{-1}U = LU = A \ . \tag{C.52}$$

The fact that the decomposition is not unique suggests the possible wisdom of forming the *normalized* decomposition

$$A = LDU \tag{C.53}$$

in which L and U are *unit* lower and upper triangular matrices and D is a diagonal matrix. The question of existence and uniqueness is addressed by Stewart (1973) who proves that the A = LDU decomposition is unique, provided the leading diagonal sub-matrices of A are nonsingular.

There are three important variants of the LDU decomposition, the first associates D with the lower triangular part to give the factorization

$$A = L^*U = (LD)U$$

where U is a unit upper diagonal matrix. This is known as the *Crout reduction* and is carried out in detail by Stewart (1973). The second variant associates D with the upper triangular factor as

$$A = LU^* = L(DU)$$

where L is a unit lower triangular matrix. This reduction is exactly that obtained by Gaussian elimination.

The third variation is possible only for symmetric positive definite matrices in which case

$$A = LDL^t \tag{C.53}$$

and the matrix $D^{\frac{1}{2}}$ is defined as

$$D^{\frac{1}{2}} = \text{diag}(d_{11}^{\frac{1}{2}}, \cdots, d_{nn}^{\frac{1}{2}}) \tag{C.54}$$

Thus A can be written as

$$A = L^*L^{*t} = (LD^{\frac{1}{2}})(D^{\frac{1}{2}}L^t) \tag{C.55}$$

where $L^* = LD^{\frac{1}{2}}$ is known as the *matrix square root*, and the factorization (C.55) is known as the *Cholesky decomposition*.

To construct an algorithm for calculating the Cholesky decomposition of a symmetric positive-definite matrix, delete the * on equation (C.55) and carry out the matrix product to verify that equating the elements yields the system of equations

$$a_{ij} = \sum_{k=1}^{i} \ell_{ik}\,\ell_{jk}\ ; \text{ for } \begin{array}{l} i = 1,2,\cdots m \\ j = 1,2,\cdots n \end{array} \tag{C.56}$$

The first few of equations (C.56) are explicitely

$$\begin{aligned} a_{11} &= \ell_{11}^2 \\ a_{12} &= a_{21} = \ell_{11}\,\ell_{21} \\ a_{13} &= a_{31} = \ell_{11}\,\ell_{31} \\ &\ \ \vdots \qquad \vdots \\ a_{1n} &= a_{n1} = \ell_{11}\,\ell_{n1} \end{aligned}$$

$$a_{22} = \ell_{22}^2 + \ell_{21}^2$$

$$a_{23} = \;_{32} = \ell_{21}\ \ell_{31} + \ell_{22}\ \ell_{32}$$

$$\vdots \qquad \vdots$$

$$a_{2n} = a_{n2} = \ell_{21}\ \ell_{n1} + \ell_{22}\ \ell_{n2}$$

$$\vdots \qquad \vdots$$

$$a_{nn} = \ell_{n1}^2 + \ell_{n2}^2 + \cdots\ \ell_{nn}^2$$

These equations can be inverted for ℓ_{ij} as follows

$$\ell_{11} = \sqrt{a_{11}}$$

$$\ell_{k1} = a_{1k}/\ell_{11} \qquad k = 2,3,\cdots n$$

$$\ell_{22} = (a_{22} - \ell_{21}^2)^{1/2}$$

$$\ell_{k2} = (a_{2k} - \ell_{21}\ \ell_{k1})/\ell_{22} \qquad k = 3,4,\cdots,n$$

Stewart (1973) summarizes these developments by the following algorithm which computes the *Cholesky decomposition* of a positive definite symmetric matrix A and returns the matrix L* in the lower half of the A matrix:

FOR $k = 1,2,\cdots n$

FOR $i = 1,2,\cdots,k-1$

$$a_{ki} \leftarrow \ell_{ki} = \ell_{ii}^{-1}(a_{ki} - \sum_{j=1}^{k-1} \ell_{ij}\ \ell_{kj}) \qquad \text{(C.57)}$$

NEXT i

$$a_{kk} \leftarrow \ell_{kk} = \left(a_{kk} - \sum_{j=1}^{k-1} \ell_{kj}^2\right)^{1/2}$$

NEXT k

If A is positive definite this algorithm can always be carried to completion. It requires about $n^3/6$ multiplications. This is about half the number of operations of Gaussian elimination; this savings is not surprising since the Cholesky decomposition takes full advantage of A's symmetry.

C.5 *Linear Least Square Elimination Algorithms*

C.5.1 *Orthogonality*

Let X and Y be non-zero 3 x 1 vectors. It is well known from analytic geometry that

$$||X|| \; ||Y|| \cos \phi = X^t Y \tag{C.58}$$

where ϕ is the angle between X and Y, and $||()||$ denotes the Eucledian norm of ().

The geometric concept of an angle does not generalize easily to higher dimensions, but by analogy with (C.58), an angle between two n x 1 vectors can be defined by the expression

$$\phi = \cos^{-1} \frac{X^t Y}{||X|| \; ||Y||} . \tag{C.59}$$

When X and Y are orthogonal the numerator of (C.59) is zero and $\phi = \pi/2$. Thus the condition $X^t Y = 0$ is said to characterize the orthogonal character of two vectors.

If two vectors X and Y are orthogonal, then

$$||X + Y||^2 = ||X||^2 + ||Y||^2 \tag{C.60}$$

This is evident since

$$\begin{aligned} ||X + Y||^2 &= (X+Y)^t (X+Y) \\ &= X^t X + Y^t Y + 2X^t Y \\ &= ||X||^2 + ||Y||^2 \end{aligned}$$

A set of n-vectors $\{U_1, U_2, \cdots, U_r\}$ are said to be an *orthonormal set* or an *orthonormal basis* if

$$U_i^t \, U_j = \delta_{ij} \tag{C.61}$$

where $\delta_{ii} = 1$ and $\delta_{ij} = 0$, $i \neq j$ is the Kronecker delta. If we define the m x n matrix $U = \{U_1 \vdots \cdots \vdots U_r\}$, then equation (C.61) can be written as

$$U^t U = I, \tag{C.62}$$

I is an r x r identity matrix. It is important to note the following important truth: For X any r x 1 matrix

$$||UX|| = ||X|| \tag{C.63}$$

This result is easily established as follows:

$$||UX||^2 = ||(UX)^t(UX)|| = ||X^t(U^tU)X|| = ||X^tX|| = ||X||^2.$$

Observe that (C.63) does not generally imply that $||X^tU|| = ||X||$; however, when U is square and satisfies (C.62), this equality is true. An *orthogonal matrix* is a square matrix satisfying $U^tU = UU^t = I$, therefore $U^{-1} = U^t$.

Theorem C.11 Every matrix A with linearly independent columns can be uniquely factored as

$$A = QR \tag{C.64}$$

where Q has orthogonal columns and R is upper triangular with positive diagonal elements.

The proof of Theorem C.11 and several algorithms for constructing the *QR factorization* are given by Stewart (1973). The most popular QR algorithm is the *Gram-Schmidt* algorithm:

Let A be an m x n matrix with linearly independent columns. Let A and Q be partitioned by columns as $A = \{A_1 \vdots \cdots \vdots A_n\}$ and $Q = \{Q_1 \vdots \cdots \vdots Q_n\}$

$Q_k = A_k$, $k = 1,2,\cdots,n$

FOR $k = 1,2,\cdots,n$

$$R_{kk} = ||Q_k||$$

$$Q_k \leftarrow R_{kk}^{-1} Q_k \tag{C.65}$$

$$R_{kj} = Q_k^t Q_j \text{ , } j = k+1,\cdots,n$$

$$Q_j \leftarrow Q_j - R_{kj} Q_k \text{ , } j = k+1,\cdots,n$$

NEXT k

We summarize another stable algorithm for constructing the QR factorization below as a by product of an *elimination procedure* for solution of the linear least squares problem.

The Gaussian elimination process developed in §C.4.2 reduces a matrix to upper triangular form via a sequence of pre-multiplications by elementary lower triangular matrices. Central to the Gaussian elimination procedure is the fact that for any given vector X, there is an elementary matrix M_k which, when pre-multiplied times X results in annihilation of the final n-k elements of X. There are other elementary matrices which can be used to introduce zeros analogous to Gaussian elimination. Here we describe an *orthogonal triangularization* process, known as the *Householder reduction* which makes use of orthogonal elementary matrices. This reduction is especially important in solution of the linear least squares problem. The elementary matrices of this elimination processes have the form

$$U = I - 2uu^t \tag{C.66}$$

where $u^t u = 1$. These matrices are known as *Householder transformations*. The fundamental properties of these elementary matrices are:

(1) U is symmetric

(2) $U^tU = I$ (U is orthogonal)

(3) $UU = I$ (U is *involutory*)

The first property is obvious. The second two properties follow immediately from

$$\begin{aligned} U^tU = U\,U &= (I-2uu^t)(I-2uu^t) \\ &= I - 4uu^t + 4uu^t uu^t \\ &= I \end{aligned}$$

In many applications, it proves advantageous to define matrices of the form (C.66) with vectors u with $u^t u \neq 1$. Letting

$2p \equiv ||u||^2 \neq 0$, it is apparent that

$$U = I - p^{-1}\, uu^t \tag{C.67}$$

is a Householder transformation.

The most important practical use of Householder transformations is their use in introducing zeros into a vector. Given a vector X, one can construct a matrix U such that $UX = -sE_1$ (the minus sign is consistent with traditional notational conventions). The scalar constant s is determined to within its sign by the fact that U is orthogonal by observing

$$s^2 = (-sE_1)^t(-sE_1) = (UX)^t(UX) = X^tX = ||X||^2 \tag{C.68}$$

so that $s = \pm\, ||X||$. Construction of u in (C.67) is based upon the following theorem.

Theorem C.12 Let X be an n x 1 matrix with $s = \pm\, ||X||$ and require that $X \neq -sE_1$. Define

$$u = X + sE_1 \tag{C.69}$$

and

$$p = \tfrac{1}{2}||u||^2 = \tfrac{1}{2}u^tu \tag{C.70}$$

then $U = I - p^{-1}\, uu^t$ is a Householder transformation with the property that $UX = -sE_1$.

The proof of this theorem follows. Since $X^tX = s^2$, then

$$\begin{aligned} p &= \tfrac{1}{2}(X + sE_1)^t\{X + sE_1\} \\ &= \tfrac{1}{2}\{X^tX + 2sx_1 + s^2\} \\ &= \tfrac{1}{2}\{2s^2 + 2sx_1\} \\ &= s^2 + sx_1 \end{aligned}$$

and therefore

$$\begin{aligned} UX &= (I - p^{-1}uu^t)X \\ &= X - p^{-1}uu^tX \\ &= X - \{(X+sE_1)(X+sE_1)^tX\}/(s^2+sx_1) \\ &= X - \{(X+sE_1)(X^tX+sx_1)\}/(s^2+sx_1) \\ &= X - \{X+sE_1\} \\ &= -sE_1 \ . \end{aligned}$$

Theorem C.12 underlies the following algorithm for computing $UX = -sE_1$. Given an n-vector $X \neq 0$, the algorithm computes s, p, and the u-vector such that $(I - p^{-1}uu^t)\ X = -sE_1$. The u-vector overwrites X:

$$\begin{aligned} & b = \max\ (|x_1|,\ |x_2|,\cdots,|x_n|) \\ & x_i \leftarrow u_i = x_i/b\ ,\ i = 1,2,\cdots,n \\ & s = \text{sign}\ (u_1)\ (u_1^2 + u_2^2 +\cdots+ u_n^2)^{\frac{1}{2}} \\ & u_1 \leftarrow u_1 + s \\ & p = s\ u_1 \\ & s \leftarrow b\ s \end{aligned} \tag{C.71}$$

Having calculated U using (C.71), one usually needs to carry out multiplication of U by some matrix A as UA. Letting $A = \{A_1 \vdots \cdots \vdots A_n\}$, then $UA = \{UA_1 \vdots \cdots \vdots UA_n\}$, so that we need consider only multipling U into a vector. Since $UA = (I - p^{-1}uu^t)A_k = A_k - (p^{-1}u^tA_k)u$, then it is not necessary to actually form U and multiply it into A numerically. The following algorithm multiplies U into a given n-vector B. The algorithm overwrites B by the product $UB = (I - p^{-1}uu^t)B$.

$$\begin{aligned} & t = p^{-1} \sum_{k=1}^{n} u_k\ b_k \\ & b_k \leftarrow b_k - tu_k \qquad k = 1,2,\cdots,n \end{aligned} \tag{C.72}$$

We now describe *Householder's reduction* process which determines a sequence of Householder transformations $U_1, U_2, \cdots, U_r$ such that the product $A_{r+1} = U_r U_{r-1} \cdots U_1 A$ is upper trapezoidal. At the kth step, $A_k = U_{k-1} U_{k-2} \cdots U_1 A$, a Householder transformation U_k is chosen to introduce zeros to all locations below the diagonal in the kth column. A_k has the form

$$A_k = \begin{pmatrix} R_k & r_k & B_k \\ 0 & c_k & D_k \end{pmatrix} \tag{C.73}$$

where R_k is an upper triangular matrix of order k-1 and c_k is a $\{m-(k-1)\}$x1 sub matrix. Using algorithm (C.71), an $\{m-(k-1)\}$x1 vector u_k and p_k can be calculated for $U_k' = I - p_k^{-1} u_k u_k^t$ so that $U_k' c_k = g_{kk} E_1$; then defining

$$U_k = \begin{matrix} I_k & 0 \\ 0 & U_k' \end{matrix} \tag{C.74}$$

then

$$A_{k+1} = U_k A_k = \begin{matrix} R_k & r_k & B_k \\ 0 & g_{kk} E_1 & U_k' D_k \end{matrix} \tag{C.75}$$

which has carried the reduction one stage further. Notice that the new upper triangular sub-matrix R_{k+1} of A_{k+1} is

$$R_{k+1} = \begin{pmatrix} R_k & r_k \\ 0 & g_{kk} \end{pmatrix} \quad . \tag{C.76}$$

The following algorithm returns the Householder transformations $U_1, U_2, \cdots, U_r$ such that $A_{r+1} = U_r U_{r-1} \cdots U_1 A$ is upper trapezoidal.

The matrices U_k are in the form $U_k = I - p_k^{-1} u_k u_k^t$, where $u_k^t = (00 \cdots 0,\ u_{k,k} u_{k+1,k} \cdots u_{m,k})$. The non-zero elements of u_k overwrite A_k. The scalars p_k are stored in $A_{m+1,k}$. R_{r+1} overwrites A_{r+1}, except the diagonal elements, which are stored in $A_{m+2,k}$.

$$
\begin{aligned}
&r = \min(m-1, n) \\
&\text{FOR } k = 1, 2, \cdots, r \\
&\quad b = \max\{|A_{ik}| : i = k, k+1, \cdots, m\} \\
&\quad \text{If } b = 0 \\
&\qquad A_{m+1,k} = 0 \\
&\qquad \text{GO TO NEXT } k \\
&\quad A_{ik} \leftarrow u_{ik} = A_{ik}/b, \quad i = k, k+1, \cdots, m \\
&\quad s = \text{sign}\,(u_{kk}) \sqrt{u_{kk}^2 + \cdots + u_{mk}^2} \\
&\quad u_{kk} = u_{kk} + \sigma \\
&\quad A_{m+1,k} \leftarrow p_k = s\, u_{kk} \\
&\quad A_{m+2,k} \leftarrow g_{kk} = -bs \\
&\quad \text{FOR } j = k+1, k+2, \cdots, n \\
&\qquad t = p_k^{-1} \sum_{i=k}^{m} u_{ik} A_{ij} \\
&\qquad A_{ij} \leftarrow A_{ij} - t\, u_{ik}, \quad i = k, k+1, \cdots, m \\
&\quad \text{NEXT } j \\
&\text{NEXT } k \\
&\text{If } m \leq n,\ A_{m+2,m} \leftarrow g_{mm} = A_{mm}
\end{aligned}
\tag{C.77}
$$

If $m > n$ (the over determined case), then this algorithm requires about $mn^2 - \frac{1}{3}n^3$ multiplications.

The most important application of the Householder reduction is solution of the linear least square problem. For $m \geq n$, the final stage of algorithm (C.77) yields

$$A_{r+1} = QA = \binom{R}{0} \tag{C.78}$$

where R is an upper triangular n x n matrix and $Q = U_r U_{r-1} \cdots U_r$. Consider the linear system

$$Y = AX + E \tag{C.79}$$

in which we desire to determine the X solution which minimizes the residual norm

$$E = Y-AX \tag{C.80}$$

If we define matrices C and D by

$$\begin{pmatrix} C \\ \cdots \\ D \end{pmatrix} = QY$$

where C is an n x 1 matrix and D is an (m-n) x 1 matrix, then observe

$$QE = Q(Y-AX) = \begin{pmatrix} C \\ \cdots \\ D \end{pmatrix} - \begin{pmatrix} RX \\ \cdots \\ 0 \end{pmatrix}$$

or

$$QE = \begin{pmatrix} C-RX \\ \cdots \\ D \end{pmatrix} \tag{C.81}$$

and, it is clear that since Q's orthogonal

$$||E||^2 = ||QE||^2 = ||C-RX||^2 + ||D||^2 \tag{C.82}$$

It is apparent from equation (C.82) that $||E||$ is minimized by choosing the X which causes the first term to be zero, viz., X is a solution of

$$RX = C \tag{C.83}$$

The importance of equations (C.83) is that R is triangular and thus X can be efficiently determined via a back substitution algorithm. The linear least squares problem can thus be solved using the Householder reduction process as follows:

(1) Use algorithm (C.77) to reduce the A matrix as $A_{r+1} = QA = \begin{pmatrix} R \\ \cdots \\ 0 \end{pmatrix}$.

(2) Calculate C and D from

$$\begin{pmatrix} C \\ \cdots \\ D \end{pmatrix} = QY$$

(3) Using the back substitution algorithm (C.20), solve the linear system (C.83) for the least square estimates.

This algorithm is more efficient and is a numerically superior means to calculate X, as compared to solution of the normal equations (1.23)

$$(A^tA)X = A^tY. \tag{C.84}$$

However, inversion of the normal equations (C.84) (using, for example, Gaussian elimination) as

$$X = (A^tA)^{-1}A^tY \tag{C.85}$$

has two possible advantages over the Householder reduction.

(1) For the case that the same A matrix is to be employed for many Y-data sets, the matrix $(A^tA)^{-1}A^t$ can be calculated *once* and simply pre-multiplied times each Y. Much less dramatic savings are possible in the Householder reduction.

(2) The matrix $(A^tA)^{-1}$ may be *required as a part of the answer* in an estimation problem, since it is the X-covariance matrix in most implementations of linear least square estimation. Thus, one may not have any choice but to attempt calculation of $(A^tA)^{-1}$, and having $(A^tA)^{-1}$ with satisfactory precision, it is much more efficient to calculate the least squares estimate from the simple matrix product in (C.85) rather than "starting over" and making use of a reduction/elimination algorithm such as algorithm (C.77).

C.6 References for Appendix C

C.1 Stewart, G.W., Introduction to Matrix Computations, Academic Press, N.Y., N.Y., ch 1, 3, 5. (1973).

C.2 Golub, G.H., "Numerical Methods for Solving Linear Least Squares Problems", Numer. Math. 7, pp 206-216 (1965).

C.3 Householder, A.S., The Approximate Solution of Matrix Problems, J. Asoc. Computing Machinery, 5, pp 204-243 (1958).

Bibliography

(1) Abramowitz, M. and L. Stegun, Handbook of Mathematical Functions, U. S. Dept. of Commerce, NBS Applied Mathematics Series, No. 55, 9th ed., 1970.

(2) Abramson, P.D., Jr., "Simultaneous Estimation of the State and Noise Statistics in Linear Dynamic Systems", Ph.D. Thesis, Massachusetts Institute of Technology, TE-25, May 1968.

(3) Aitken, A.C., "On Least Squares and Linear Combinations of Observations", Proc. Roy. Soc. Edinburgh, Sect. A, 55 (1955), 42-48.

(4) Albert, A. and R.W. Sittler, "A Method for Computing Least Squares Estimators That Keep Up with the Data", J. SIAM Control, ser. A, vol. 3, pp. 384-417, 1965.

(5) Albert, A.E. and Gardner, L.A., Jr., Stochastic Approximation and Nonlinear Regression, The M.I.T. Press, Cambridge, Mass. 1967.

(6) Anderson, B.D.O., "An Algebraic Solution to the Spectral Factorization Problem", IEEE Trans. Auto. Control, vol. AC-12, no. 4, pp. 410-414, August 1967.

(7) Anderson, B.D.O., and J.B. Moore, "State Estimation via the Whitening Filter", Proc. Joint Autom. Control Conf., June 1968, pp. 123-129, Ann Arbor, Mich.

(8) Anderson, T. W., An Introduction to Multivariate Statistical Analysis, John Wiley and Sons, Inc., New York, 1958.

(9) Andrews, A., "A Square Root Formulation of the Kalman Covariance Equations", AIAA Journal, vol. 6, no. 6, pp. 1165-1166, 1968.

(10) Aoki, M. and J.R. Huddle, "Estimation of the State Vector of a Linear Stochastic System with a Constrained Estimator", IEEE Trans. Autom. Control, vol. AC-12, no. 4, pp. 432-433, August, 1967.

(11) Athans, M., "The Relationship of Alternate State-Space Representations in Linear Filtering Problems", IEEE Trans. Autom. Control, vol. AC-12, no. 6, pp. 775-776, December 1967.

(12) Athans, M., and E. Tse, "A Direct Derivation of the Optimal Linear Filter Using the Maximum Principle", IEEE Trans. Autom. Control, vol. Ac-12, no. 6, pp. 690-698, December 1967.

(13) Athans, M. and R.P. Wishner and A. Bertolini, "Suboptimal State Estimation for Continuous-time Nonlinear Systems from Discrete Noisy Measurements", IEEE Trans. Autom. Control, vol. AC-13, no. 5, October 1968, pp. 504-514.

(14) Baggeroer, A.B., "Maximum a Posteriori Interval Estimation", presented at WESCON, August 1966, session 7, The Application of State-variable Techniques to Communication and Radar Problems.

(15) Baghdady, E.J. and K.W. Kruse, "Signal Design for Space Communication and Tracking Systems", IEEE Trans. Commun. Technol., vol. COM-13, no. 4, pp. 484-498, December 1965.

(16) Balakrishnan, A.V., "A General Theory of Nonlinear Estimation Problems in Control Systems", J. Math. Anal. Appl., vol. 8, no. 1, pp. 4-30, February 1964.

(17) Balakrishnan, A.V., "On a Class of Nonlinear Estimation Problems", IEEE Trans. Inform. Theory, vol. IT-10, no. 4, pp. 314-320, 1964.

(18) Bass, R.W., V.D. Norum and L. Schwartz, "Optimal Multi-channel Nonlinear Filtering", J. Math. Anal. Appl., vol 16, pp. 152-164, 1966.

(19) Battin, R.H., "A Statistical Optimizing Navigation Procedure for Space Flight", ARS J, vol. 32, pp. 1681-1696, 1962.

(20) Battin, R.H., "Astronautical Guidance, McGraw-Hill Book Company, New York, 1964.

(21) Beckman, F.S., "The Solution of Linear Equations by the Conjugate Gradient Method", Mathematical Methods for Digital Computers, vol. 1, A. Ralson and H. Wilf(eds.), New York: Wiley, 1960.

(22) Beckmann, P., Probability in Communication Engineering, Harcourt Brace & World, Inc., New York, 1967.

(23) Bellantoni, J.F. and K.W. Dodge, "A Square Root Formulation of the Kalman-Schmidt Filter", AIAA J., vol. 5, no. 7, pp. 1309-1314, July 1967.

(24) Bellman, R.E., H.H. Kagiwada, R.E. Kalaba and R. Sridhar, "Invariant Imbedding and Nonlinear Filtering Theory", J. Astron. Sci., vol. 13, pp. 110-115, May-June 1966.

(25) Bellman, R.E. and R. Kalaba, "On the Fundamental Equations of Invariant Imbedding-I", Proc. National Academy of Science, U.S.A., vol. 47, pp. 336-338, 1961.

(26) Bello, P., "Joint Estimation of Delay, Doppler, and Doppler rate", IRE, Trans. PGIT, IT-6, no. 3, pp. 330-341, June 1960.

(27) Bendat, J.S., "A General Theory of Linear Prediction and Filtering", J. Soc. Indust, Appl. Math., 4, pp. 131-151, Sept. 1956.

(28) Bendat, J.S., "Exact Integral Equation Solutions and Synthesis for a Large Class of Optimum Time Variable Linear Filters", IRE, Trans. PGIT, IT-3, pp. 71-80, March 1957.

(29) Bendat, J.S., Principles and Application of Random Noise Theory, John Wiley & Sons, Inc., New York, 1958.

(30) Bendat, J.S. and A.G. Pierson, Measurement and Analysis of Random Data, John Wiley & Sons, Inc., New York, 1966.

(31) Benedict, T.R. and M.M. Sondhi, "On a Property of Wiener Filters", Proc. IRE, 45 (July, 1957), 1021-1022.

(32) Berkson, J., "Estimation by Least Squares and by Maximum Likelihood", Proceedings of the Third Berkeley Symposium on Mathematical Statistics and Probability, Berkeley, Calif., University of California Press, 1956, Vol. I, 1-11.

(33) Bharucha-Reid, A.T., Elements of the Theory of Markov Processes and Their Applications, McGraw-Hill Book Company, New York, 1960.

(34) Bierman, C.J., "Measurement Updating Using the U-D Factorization", Proceedings 1975 IEEE Conference on Decision and Control, pp. 337-346, 1975.

(35) Bizwas, K.K. and A.K. Mahalanabis, "Suboptimal Algorithms for Nonlinear Smoothing", IEEE Trans. on Aerospace and Electronic Systems, Vol. AES-9, No. 4, pp. 529-534, July 1973.

(36) Bjerhammar, A., "Rectangular Reciprocal Matrices, with Special Reference to Geodetic Calculations", Bull. Geod. Int., 52 (1951), 188-220.

(37) Bjork, A., "Solving Linear Least Squares Problems by Gram-Schmidt Orthogonalization", BIT 7,(1967) 1-21.

(38) Bjork, A., "Iterative Refinement of Linear Least Squares Solution I", BIT 7, (1967), 251-278.

(39) Bjork, A. and G.H. Golub, "Iterative Refinement of Linear Least Squares Solution II", BIT 8, (1968), 8-30.

(40) Bjork, A. and G.H. Golub, "Iterative Refinement of Linear Least Squares Solution by Householder Transformation", BIT 7, (1967), 322-337.

(41) Blackman, R.B. and J.W. Tukey, The Measurement of Power Spectra from the Point of View of Communications Engineering, New York, Dover Publications, Inc., 1959.

(42) Blackman, R.B., "Methods of Orbit Refinement", Bell System Tech. J., 43 (May 1964), 885-909.

(43) Blanton, J.N. and J.L. Junkins, "Dynamical Constraints in Satellite Photogrammetry", AIAA Paper No. 76-824 presented to the AIAA/AAS Astrodynamics Specialists Conference, San Diego, Ca., August 1976.

(44) Blanton, J.N., "A New Formulation of the Free Motion of Triaxial Rigid Bodies and Applications to Satellite Photogrammetric Triangulation", Ph.D. Dissertation, Univ. of Va., Charlottesville, Va., August 1976.

(45) Blanton, J.N. and J.L. Junkins, "Dynamical Constraints in Satellite Photogrammetry", AIAA Paper #76-824, AIAA/AAS Astrodynamics Conference, San Diego, Ca., August 1976, to appear in AIAA Journal, May 1977.

(46) Blum, M., "A Stagewise Parameter Estimation Procedure for Correlated Data", Numer. Math., Vol. 3, 1961, pp. 202-208.

(47) Bode, H.W. and C.E. Shannon, "A Simplified Derivation of Linear Least-squares Smoothing and Prediction Theory", Proc. IRE, vol. 38, pp. 417-425, 1950.

(48) Bona, B.E. and R.J. Smay, "Optimum Reset of Ship's Inertial Navigation System", IEEE Transactions on Aerospace and Electronic Systems, Vol. AES-2, No. 4, July 1966, pp. 409-414.

(49) Bongiorno, J.J., Jr. and D.C. Youla, "On Observers in Multivariable Control Systems", Intern. J. Control, vol. 8, no. 3, pp. 221-243, 1968.

(50) Booton, R.C., "An Optimization Theory for Time-varying Linear Systems with Non-stationary Statistical Inputs", Proc. IRE, vol. 40, pp. 977-981, 1952.

(51) Britting, K.R., Inertial Navigation System Analysis, John Wiley & Sons, New York, 1971.

(52) Brown, D. "The Rigorous and Simultaneous Adjustment of Lunar Orbital Photography", Rome Air Development Center Report RADC-TR-70-274, 1970.

(53) Brown, D., "A Unified Lunar Control Network", Photogrammetric Engineering, 34:12, pp. 1272-1292, 1968.

(54) Brown, R.G., "Analysis of an Integrated Inertial-Doppler-Satellite Navigation System, Part 1, Theory and Mathematical Models", Engineering Research Institute, Iowa State University, Ames, Iowa, ERI-62600, November 1969.

(55) Brown, R.G. and G.L. Hartmann, "Kalman Filter with Delayed States as Observables", Proceedings of the National Electronics Conference, Chicago, Illinois, Vol. 24, December 1966. pp. 67-72.

(56) Bryson, A. and Y. Ho, Applied Optimal Control, Blaisdell, London, Ch 10 (1969).

(57) Bryson, A.E., Jr., "Applications of Optimal Control Theory in Aerospace Engineering", J. Spacecraft Rockets, vol. 4, pp. 545-553, May 1967.

(58) Bryson, A.E., Jr. and L.J. Henrikson, "Estimation Using Sampled-data Containing Sequentially Correlated Noise", Harvard University, Div. Eng. Appl. Phys., Tech. Rept. 533, Cambridge, Mass., June 1967.

(59) Bryson, A.E., Jr. and D.E. Johansen, "Linear Filtering for Time-varying Systems Using Measurements Containing Colored Noise", IEEE Trans. Autom. Control, vol. AC-10, no. 1, pp. 4-10, January 1965.

(60) Bryson, A.E., Jr. and M. Frazier, "Smoothing for Linear and Non-linear Dynamic Systems", Proc. Optimum Sys. Synthesis Conf., U.S. Air Force Tech. Rept. ASD-TDR-63-119, Feb. 1963.

(61) Bucy, R.S., "Nonlinear Filtering Theory", IEEE Trans. Autom. Control, vol. AC-10, no. 2, pp. 198-199, April 1965.

(62) Bucy, R.S., "Optimal Filtering for Correlated Noise", J. Math. Anal. Appl., vol. 20, pp. 1-8, 1967.

(63) Bucy, R. S. and P.D. Joseph, Filtering for Stochastic Processes with Applications to Guidance, John Wiley & Sons, Inc., New York, 1968.

(64) Businger, P.A. and G.H. Golub, "Linear Least Squares Solutions by Householder Transformations", Numer. Math. 7, 269-276, 1965.

(65) Businger, P. A. and G.H.Golub, "Algorithm 358, Singular Value Decomposition of a Complex Matrix", Comm. ACM 12, 564-565, 1969.

(66) Cameron, J.M., "An Algorithm for Obtaining an Orthogonal Set of Individual Degrees of Freedom for Error", J. Res. Nat. Bur. Std., 67B, pp. 19-22, Jan-Mar 1961.

(67) Carlson, N.A., "Fast Triangular Formulation of the Square Root Filter", AIAA Journal, Vol. 11, No. 9, 1973, pp. 1259-1264.

(68) Carlton, A.G. and J.W. Follin, Jr., "Recent Developments in Fixed and Adaptive Filtering", Proc. of the Second AGARD Guided Missiles Seminar (Guidance and Control), AGARD Monograph 21, (Sept. 1956).

(69) Carney, T.M., "On Joint Estimation of the State and Parameters for an Autonomous Linear System Based on Measurements Containing Noise", Ph.D. Thesis, Rice University, 1967.

(70) Carney, T.M. and R.M. Goldwyn, "Numerical Experiments with Various Optimal Estimators", J. Opt. Theory Appl., vol. 1, no. 2, pp. 113-130, 1967.

(71) Chang, S.S.L., Synthesis of Optimum Control Systems, McGraw-Hill Book Company, New York, 1961.

(72) Chauvenet, W., A Manual of Spherical and Practical Astronomy: Vol. I, Spherical Astronomy; Vol. II, Theory and Use of Astronomical Instruments Method of Least Squares, Dover Publications, Inc., New York, 1960 (reprint).

(73) Chen, R.T.N., "A Recurrence Relationship for Parameter Estimation by the Method Quasilinearization and Its Connection with Kalman Filtering", Joint Automatic Control Conference, Atlanta, Georgia, June 1970.

(74) Chernoff, H. and L.E. Moses, Elementary Decision Theory, John Wiley and Sons, Inc., New York, 1959.

(75) Chin, P.P., "Real Time Kalman Filtering of APOLLO LM/AGS Rendezvous Radar Data", AIAA Guidance Control and Flight Mechanics Conference, Santa Barbara, California, August 1970.

(76) Choate, W.C. and A.P. Sage, "Operator Algebra for Differential Systems", IEEE Trans. System Sci. and Cybernetics, vol.SSC-3, no. 3, pp. 137-147, December 1967.

(77) Claus, A.J., R.B. Blackman, E.G. Halline and W.C. Ridgway,III, "Orbit Determination and Prediction and Computer Programs", Bell System Tech. J., 42, pp. 1357-1382, July 1963.

(78) Cohen, C., "Lectures on Least Squares", edited by T.M. Alexander, U.S. Naval Weapons Laboratory Report KGR-M1, Dahlgren, Va., May 1972.

(79) Cole, J.D., Perturbation Methods in Applied Mathematics, Wiley, ch. 1-4, 1968.

(80) Collins, L.D., "Realizable Whitening Filters and State-variable Realizations", Proc. IEEE, vol. 56, no. 1, pp. 100-101, January, 1968.

(81) Collmeyer, A.J. and S.C. Gupta, "Linear Filtering with Constraints, Proc. Joint Autom Control Conf, Ann Arbor, Mich., pp. 137-144, June 1968.

(82) Cox, D.R. and H.D. Miller, Theory of Stochastic Processes, John Wiley & Sons, Inc., New York, 1965.

(83) Cox, H., "On the Estimation of State Variables and Parameters for Noisy Dynamic Systems", IEEE Trans. Autom. Control, vol. AC-9, no. 1, pp. 5-12, January 1964.

(84) Cox, H., "Estimation of State Variables via Dynamic Programming", Proc. Joint Autom. Control Conf., Stanford Univ., Leland-Stanford, Calif., pp. 376-381, June 1964.

(85) Cramer, H., Mathematical Methods of Statistics, Princeton University Press, Princeton, N.J., 1946.

(86) D'Appolito, J.A. and C.E. Hutchinson, "Low Sensitivity Filters for State Estimation in the Presence of Large Parameter Uncertainties", IEEE Trans. on Automatic Control, Vol. AC-14, No. 3, pp. 310-312, June 1969.

(87) D'Appolito, J.A., "The Evaluation of Kalman Filter Designs for Multisensor Integrated Navigation Systems", The Analytic Sciences Corp., AFAL-TR-70-271, (AD 881286), January 1971.

(88) D'Appolito, J.A., "A Simple Algorithm for Discretizing Linear Stationary Continuous Time Systems", Proc. of the IEEE, Vol. 54, No. 12, pp. 2010-2011, December 1966.

(89) Darlington, S., "Linear Least-Squares Smoothing and Prediction, with Applications", Bell System Tech. J., 37, pp. 1221-1293, Sept. 1958.

(90) Darlington, S., "Nonstationary Smoothing and Prediction Using Network Theory Concepts", IRE, Trans. PGIT, IT-5, Special Supplement, 1-13, May 1959.

(91) Davenport, W.B. and W.L. Root, Introduction to Random Signals and Noise, McGraw-Hill Book Company, New York, 1958.

(92) Davis, M.C. "Factoring the Spectral Matrix", IEEE Trans. Autom. Control, vol. AC-8, no. 4, pp. 296-305, October 1963.

(93) De Groot, M.H. and M.M. Rao, "Bayes Estimation with Convex Loss", Annals Math. Stat., 34 pp. 839-846, September 1963.

(94) Denham, W.F. and S. Pines, "Sequential Estimation When Measurement Function Nonlinearity Is Comparable to Measurement Error", AIAA J., vol. 4, no. 6, pp. 1071-1076, June 1966.

(95) Derman, C., "Stochastic Approximation", Ann. Math. Statistics, Vol. 27, pp. 879-886, 1956.

(96) Detchmendy, D.M. and R. Sridhar, "Sequential Estimation of States and Parameters in Noisy Nonlinear Dynamical Systems", Trans. ASME, J. Basic Eng., vol. 88D, pp. 362-366, April 1966.

(97) Deutsch, R., Estimation Theory, Englewood Cliffs, N.J., Prentice-Hall, 1965.

(98) Deutsch, R., Nonlinear Transformation of Random Processes, Englewood Cliffs, N.J., Prentice-Hall, 1962.

(99) Deyst, J.J., "A Derivation of the Optimum Continuous Estimator for Systems with Correlated Measurement Noise", AIAA Journal, Vol. 7, No. 11, September 1969.

(100) Deyst, J.J. and C.F. Price, "Conditions for Asymptotic Stability of the Discrete, Minimum Variance, Linear Estimator", IEEE Trans. on Automatic Control, Vol. AC-13, No. 6, Dec. 1968, pp. 702-705.

(101) Doob, J.L., Stochastic Processes, John Wiley & Sons, Inc., New York, 1953.

(102) Downey, J.J., S. Hollander and E.A. Rottach, "A Loran Inertial Navigation System", Sperry Engineering Review, Vol. 21, No. 1, pp. 36-43, 1968.

(103) Dubes, R.C., The Theory of Applied Probability, Prentice-Hall, Inc., Englewood Cliffs, N.J., 1968.

(104) Dyer, P. and S. McReynolds, "Extension of Square-Root Filtering to Include Process Noise", J. Opt. Theory Appl., vol. 3, no. 6, pp. 444-458, 1969.

(105) Edgeworth, F.Y., "On the probable error of frequency-constants", Jour. Roy. Stat. Soc., 71, pp. 381-397, 1908.

(106) Esposito, R., "On a Relation between Detection and Estimation in Decision Theory", Inform. Control, vol. 12, pp. 116-120, 1968.

(107) Eykhoff, P., System Identification: Parameter and State Estimation, Wiley, New York, 1974.

(108) Fagin, S.L., "Recursive Linear Regression Theory, Optimal Filter Theory, and Error Analyses of Optimal Systems", 1964 IEEE Intern. Conv. Record, New York, pp. 216-240, March 1964.

(109) Fagin, S.L., "Feedback Realization of a Continuous-time Optimal Filter", IEEE Trans. Aerospace Electron Syst., vol. AES-3, no. 3, pp. 494-509, May 1967.

(110) Fagin, S.L., "The Weighting Function Form of the Continuous Optimal Filter", Proc. Joint Autom. Control Conf., June 1968, pp. 619-625, Ann Arbor, Mich.

(111) Fagin, S.L., E. Grinoch and A. Graefe, "Continuous Time-varying Optimal Feedback Applied to the Augmentation and Rapid Alighment of Inertial Systems", IEEE Aerospace Systems Conf., Seattle, Wash., pp. 661-678, July 1966.

(112) Fang, B.T., "Nonlinear Counter Example for Batch and Extended Sequential Estimation Algorithms", Flight Mechanics/Estimation Theory Symposium, NASA GSFC preprint X-582-75-273, Greenbelt, Md., August 1975.

(113) Fang, B.T., "Kalman-Bucy Filter for Optimum Radio-inertial Navigation", IEEE Trans. Autom. Control, vol AC-12, no. 4, pp. 430-431, August 1967.

(114) Feller, W., An Introduction to Probability Theory and Its Applications, vol. I, John Wiley & Sons, Inc., New York, 1950, 1957.

(115) Feller, W., An Introduction to Probability Theory and Its Applications, vol. II, John Wiley & Sons, Inc., New York, 1966.

(116) Fiacco, A.V. and G.P. McCormick, Nonlinear Programming: Sequential Unconstrained Minimization Techniques, Wiley, New York, Ch. 8, 1968.

(117) Fisher, J.R., "Optimal Nonlinear Filtering", Advan. Control Systems, vol. 5, pp. 197-300, 1967.

(118) Fisher, J.R. and E.B. Stear, "Optimal Nonlinear Filtering for Independent Increment Processes, Parts I and II", IEEE Trans. Information Theory, Vol. IT-13, No. 4, pp. 558-578.

(119) Fisher, R.A., "On an absolute criterion for Fitting Frequency Curves", Mess. of Math., 41, pp. 155, 1912.

(120) Fisher, R.A., "Theory of Statistical Estimation", Proc. Cambridge Phil. Soc., vol. 22, pp. 700, 1925.

(121) Fitzgerald, R.J., "Divergence of the Kalman Filter", IEEE Trans. Auto. Control, Vol. AC-16, no. 6, pp. 736-747, December 1971.

(122) Fitzgerald, R.J., "Error Divergence in Optimal Filtering Problem", Second IFAC Symposium on Automatic Control in Space, Vienna, Austria, September 1967.

(123) Fletcher, R. and M.J.D. Powell, "A Rapidly Convergent Descent Method for Minimization", Computer Journal 6, pp. 163-168, 1963.

(124) Fletcher, R. and C.M. Reeves, "Function Minimization by Conjugate Gradients", Computer Journal 7, pp. 149-154, 1964.

(125) Fortmann, T., D.L. Kleinman, and M. Athans, "On the Design of Linear Systems with Piecewise-Constant Feedback Gains", IEEE Trans. on Automatic Control, Vol. AC-13, No. 4, pp. 354-361, August 1968.

(126) Foster, M., "An Application of the Wiener-Kolmogorov Smoothing Theory to Matrix Inversion", J. Soc. Indust. Appl. Math., 9, no. 3, pp. 387-392, Sept. 1961.

(127) Fraser, Donald C., "On the Application of Optimal Linear Smoothing Techniques to Linear and Nonlinear Dynamic Systems", Ph.D. Thesis, Massachusetts Institute of Tech., January 1967.

(128) Freund, R.J., R.W. Vail and C.W. Clunies-Ross, "Residual Analysis", J. Amer. Statist. Assoc., 56, pp. 98-104, 1961.

(129) Friedland, B., "Least Squares Filtering and Prediction of Nonstationary Sampled Data", Information and Control, I pp. 297-313, 1958.

(130) Friedland, B., "Treatment of Bias in Recursive Filtering", IEEE Trans. Autom. Control, vol. AC-14, no. 4, pp. 359-367, Aug. 1969.

(131) Friedland, B. and I. Bernstein, "Estimation of the State of a Nonlinear Process in the Presence of Nongaussian Noise and Disturbances", J. Franklin Inst., vol. 281, no. 6. pp. 455-480, June 1966.

(132) Frost, P.A., "Nonlinear Estimation in Continuous Time Systems", Stanford Univ., Systems Theory Lab., Tech. Rept. 6304-4, U.S. Air Force Contract F3361567-C-1245, May 1968.

(133) Gantmacher, F.R., The Theory of Matrices, vols. I and II, Chelsea Publishing Company, New York, 1959.

(134) Gauss, K.F., Theory of the Motion of the Heavenly Bodies Moving About the Sun in Conic Sections, Dover Publications, Inc., reprinted 1963.

(135) Gelb, A. (editor), Applied Optimal Estimation, the MIT Press, Cambridge, Mass., 1974.

(136) Gelb, A., A. Dushman, and H.J. Sandberg, "A Means for Optimum Signal Identification", NEREM Record, Boston, November 1963.

(137) Gelb, A. and W. E. Vander Velde, Multiple-Input Describing Functions and Nonlinear System Design, McGraw-Hill Book Co., Inc., New York, 1968.

(138) Gelb, A. and R.S. Warren, "Direct Statistical Analysis of Nonlinear Systems", Proc AIAA Guidance and Control Conf., Palo Alto, August 1972.

(139) Gentleman, W.M., "Least Square Computations by Givens Transformations Without Square Roots", J. Inst. Maths. Applic., Vol. 12, pp. 329-336, 1973.

(140) Goldberger, A.S., "Note on Stepwise Least Squares", J. Amer. Statist. Assoc., 56, pp. 105-110, 1961.

(141) Goldberger, A.S., "Stepwise Least Squares: Residual Analysis and Specification Error", J. Amer. Statist. Assoc., 56, pp. 998-1000, 1961.

(142) Golonb, S.W., Digital Communications with Space Applications, Prentice-Hall, Inc., Englewood Cliffs, N.J., 1964.

(143) Golub, G.H., "Comparison of the Variance of Minimum Variance and Weighted Least Squares Regression Coefficients", Ann. Math. Statist., 34, pp. 984-991, Sept. 1963.

(144) Golub, G.H., "Numerical Methods of Solving Linear Least Squares Problems", Numer. Math. 7, pp. 206-216, 1965.

(145) Golub, G.H. and C. Reinsch, "Singular Value Decomposition and Least Squares Solutions", Numer. Math. 14, pp. 403-420. 1970.

(146) Golub, G.H. and M.A. Sanders, "Linear Least Squares and Quadratic Programmin", Tech. Rept. No. CS134. Comput. Sci. Dept., Stanford Univ., Stanford, Calif., 1969.

(147) Golub, G.H. and J.H. Wilkinson, "Note on Iterative Refinement of Least Squares Solution", Numer. Math. 9, pp. 139-148, 1966.

(148) Gray, A.H. and T.K. Caughey, "A Controversy in Problems Involving Random Parametric Excitation", J. Math. Phys., vol. 44, pp. 288-296, September 1965.

(149) Graybill, F. A., An Introduction to Linear Statistical Models, vol. 1, McGraw-Hill Book Company, New York, 1961.

(150) Graybill, F. A. and R. B. Deal, "Combining Unbiased Estimators", Biometrics, 15, pp. 543-550, 1959.

(151) Grenander, U. and M. Rosenblatt, Statistical Analysis of Stationary Time Series, John Wiley & Sons, Inc., 1957.

(152) Greville, T.N.E., "The Pseudoinverse of a Rectangular or Singular Matrix and its Application to the Solution of Systems of Linear Equations," SIAM Rev., 1, pp. 38-43, 1959.

(153) Greville, T.N.E., "Some Applications of the Pseudoinverse of a Matrix", SIAM Rev., 2, pp. 15-22, 1960.

(154) Griffin, R.E. and A.P. Sage, "Large and Small Scale Sensitivity Analysis of Optimum Estimation Algorithms", IEEE Trans. Autom. Control, vol. AC-13, no. 4, pp. 320-329, August 1968.

(155) Griffin, R.E. and A.P. Sage, "Error Bounds for Optimum Smoothing Algorith-s", Proc. Natl. Electron. Conf., pp. 58-63, December 1968.

(156) Griffin, R.E. and A.P. Sage, "Sensitivity Analysis of Fixed Point Linear Smoothing Algorithms", Intern J. Control, vol. 8, no. 4, pp. 321-337, 1968.

(157) Griffin, R.E. and A.P. Sage, "Error Bounds in Optimum Smoothing, Proc. IEEE, vol. 57, no. 4, pp. 725-726, April 1969.

(158) Griffin, R.E. and A.P. Sage, "Sensitivity Analysis of Discrete Filtering and Smoothing Algorithms", AIAA J., vol. 7, no. 10, pp. 1890-1897, October 1969.

(159) Gura, I.A., "An Algebraic Approach to Optimal Estimation", Hughes Space Systems Division Report No. SSD 70072R, El Segundo, California, 1967.

(160) Hancock, J.C. and P.A. Wintz, Signal Dectection Theory, McGraw-Hill Book Company, New York, 1966.

(161) Hanson, J.E., "Some Notes on the Application of the Calculus of Variations to Smoothing for Finite Time, etc.", JHU/APL Internal Memorandum, BBD-346, 1957.

(162) Helstrom, C.W., Statistical Theory of Signal Detection, Pergamon Press, Inc., London, 1960.

(163) Henrikson, L.J., "Sequentially Correlated Measurement Noise with Application to Inertial Navigation", Ph.D. Thesis, Harvard University, 1967.

(164) Herrick, S., Astrodynamics, Vol. II, Van Nostrand Reinhold 1971.

(165) Ho, B.L. and R. E. Kalman, "Spectral Factorization Using the Riccati Equation", Aerospace Corp. Rept. TR-1001(2307)-1, November 1966.

(166) Ho, Y.C., "The Method of Least Squares and Optimal Filtering Theory", RAND Corp. RM-3329-PR, October 1963.

(167) Ho, Y.C. and A.K. Agrawala, "On Pattern Classification Algorithms, Introduction and Survey", Proc. IEEE, vol. 56, no. 12, pp. 2101-2114, December 1968.

(168) Ho, Y.C. and R.C.K. Lee, "A Bayesian Approach to Problems in Stochastic Estimation and Control, "IEEE Trans. Autom. Control, vol. AC-9, no. 4, pp. 333-339, October 1964.

(169) Ho, Y.C., "On the Stochastic Approximation Method and Optimal Filter Theory", J. Math. Anal. Appl., vol. 6, pp. 152-154, 1962.

(170) Holtzman, J.M., "Signal-Noise Ratio Maximization Using the Pontryagin Maximum Principle", Bell System Tech. J., vol. 45, no. 3, pp. 473-488, March 1966.

(171) Householder, A.S. and G. Young, "Matrix Approximation and Latent Roots", Amer. Math. Monthly, 45, pp. 165-171, 1938.

(172) Householder, A.S., "The Approximate Solution of Matrix Problems", J. Assoc. Comput. Mach. 5, pp. 204-243, 1958.

(173) Householder, A.S., "Unitary Triangularization of a Non-symmetric Matrix", J. Assoc. Comput. Mach. 5, pp. 339-342. 1958.

(174) Householder, A.S., The Theory of Matrices in Numerical Analysis, Ginn(Blaisdell), Boston, Massachusetts, 1964.

(175) Householder, A.S., and F. L. Bauer, "On Certain Methods for Expanding the Characteristic Polynomial", Numer. Math.1, pp. 29-37. 1958.

(176) Huddle, J.R., "On Suboptimal Linear Filter Design", IEEE Trans. on Automatic Control, Vol. AC-12, No. 3, pp. 317-318, June 1967.

(177) Hutchinson, C.E. and J.A. D'Appolito, "Design of Low-Sensitivity Kalman Filters for Hybrid Navigation Systems", AGARD Conference Proceedings No. 54 on Hybrid Navigation Systems, AGARD CP No. 54, January 1970.

(178) Hutchinson, C.E. and R.A. Nash, Jr., "Comparison of Error Propagation in Local-Level and Space-Stable Inertial Systems", IEEE Transactions on Aerospace and Electronic Systems, Vol. AES-7, No.6, pp. 1138-1142, November 1971.

(179) Huzurbazar, V.S., "The Likelihood Equation, Consistency and the Maxima of the Likelihood Function", Ann. Eugen., Lond., 14, pp. 185-200, 1948.

(180) Ince, E.L., Ordinary Differential Equations, Longmans, London, pp. 540-547, 1926.

(181) Jacobson, D.H., "Second-Order and Second Variation Methods for Determining Optimal Control: A Comparative Analysis Using Differential Dynamic Programming", Intl. J. of Control, Vol. 7, No. 2, pp. 175-196, 1968.

(182) Jazwinski, A.H., "Limited Memory Optimal Filtering", IEEE Trans. Autom. Control, vol. AC-13, no. 5, pp. 558-563, October 1968.

(183) Jazwinski, A.H., "Adaptive Filtering", Automatica, vol. 5, no. 4, pp. 475-485, 1969.

(184) Jazwinski, A.H., Stochastic Processes and Filtering Theory, Academic Press, N.Y., ch. 8 and 9, 1970.

(185) Johansen, D.E., "Solution of a Linear Mean Square Estimation Problem When Process Statistics Are Undefined", IEEE Trans. Autom. Control, vol. AC-11, no. 1, pp. 20-30, January 1966.

(186) Junkins, J.L., "Computing Variant Powered Trajectories Using Encke's Method", McDonnell Douglas Astronautics Co. Memorandum A2-260-AAC5-D7480, September 1967.

(187) Junkins, J.L., "On the Optimization and Estimation of Powered Rocket Trajectories ...", McDonnell Douglas Astronautics Co., Santa Monica, Calif., Report No. SM G1793, 1969.

(188) Junkins, J.L., "Equivalence of the Minimum Norm and Gradient Projection Constrained Optimization Techniques", AIAA J., Vol. 10, No. 7, pp. 927-929, July 1972.

(189) Junkins, J.L., "Estimation Theory", Naval Surface Weapons Center Report No. TN-K-45h4, Dahlgren, Va., January 1975.

(190) Junkins, J.L., "Development of a Finite Element Gravity Field", Research Laboratories for the Engineering Sciences, Report No. ESS-3538-105-75, Univ. of Va., Charlottesville, Va., December 1975.

(191) Junkins, J.L., "An Investigation of Finite Element Representation of the Geopotential", AIAA J., vol. 14, no. 6, pp. 803-808, June 1976.

(192) Junkins, J.L., "Development of Finite Element Models for the Earth's Gravity Field: Macro Model for Satellite Orbit Integration", Research Laboratories for the Engineering Sciences Report No. UVA/525023/ESS77/103, Univ. of Va., Charlottesville, Va., February 1977.

(193) Junkins, J.L., "Development of Finite Element Models for the Earth's Gravity Field: Fine Structure Model for Inertial Guidance", Research Laboratories for the Engineering Sciences Report No. UVA/525023/ESS77/104, Univ. of Va., Charlottesville, Va., February 1977.

(194) Junkins, J.L. and J.N. Blanton, "New Analytical Methods for Simulation/Estimation of Asymmetric Satellite Dynamics", Research Laboratories for the Engineering Sciences Report No. ESS-3323-101-73, Univ. of Va., Charlottesville, Va., December 1973.

(195) Junkins, J.L. and J.R. Jancaitis, "Weighting Function Techniques for Storage and Analysis of Mass Remote Sensing Data", Machine Processing of Remotely Sensed Data, published by Laboratory for Applications of Remote Sensing, Purdue Univ., West Lafayette, Ind., October 1973.

(196) Junkins, J.L. and J.R. Jancaitis, "Piecewise Continuous Surfaces: A Flexible Basis for Digital Mapping", Proceedings of the American Society of Photogrammetry, pp. 1-18, September 1974.

(197) Junkins, J.L., J.R. Jancaitis, and G.W. Miller, "Smooth Irregular Curves, Photogrammetric Engr., vol. 38, no. 6, pp. 565-573, June 1972.

(198) Junkins, J.L., G.W. Miller, and J.R. Jancaitis, "A Weighting Function Approach to Modeling of Irregular Surfaces", Journal of Geophysical Research, vol. 78, no. 11, pp. 1794-1803, April 1973.

(199) Junkins, J.L., C.C. White III, and J.D. Turner, "A New System for Star Pattern Recognition/Spacecraft Attitude Determination", presented to the Flight Mechanics/Estimation Theory Symposium held at NASA Goddard Space Flight Center, Greenbelt, MD., October 27-28, 1976, proceedings in press.

(200) Kac, M. and A.J.F. Siegert, "An Explicit Representation of a Stationary Gaussian Process", Ann. Math. Statist., 18, pp. 38, 1947.

(201) Kagiwada, H.H., R.E. Kalaba, A. Schumitzky, and R. Sridhar, "Invariant Imbedding and Sequential Interpolating Filters for Nonlinear Processes", Trans. ASME, J. Basic Eng., vol. 91D, no. 2, pp. 195-200, June 1969.

(202) Kahan, W. and B.N. Parlett, "On the Convergence of a Practical QR Algorithm", Proc. IFIP Congr. A25-A30, 1968.

(203) Kailath, T., "Adaptive Matched Filters, "Proc. Symp. Math. Optimization Tech., University of Calif Press, Berkeley, 1963.

(204) Kailath, T., "An Innovations Approach to Least Squares Estimation-Part I: Linear Filtering in Additive White Noise", IEEE Trans. Autom. Control, vol. AC-13, no. 6, pp. 646-655, Dec. 1968.

(205) Kailath, T., "A General Likelihood Ratio Formula for Random Signals in Gaussian Noise", IEEE Trans. Inform. Theory, vol. IT-15, no. 3, pp. 350-361, May 1969.

(206) Kailath, T. and P. Frost, "Mathematical Modeling for Stochastic Process", Stochastic Problems in Control, Proc. of Symposium of AACC, published by ASME, Ann Arbor, Mich., June 1968.

(207) Kailath, T. and P. Frost, "An Innovations Approach to Least Squares Estimation-Part II: Linear Smoothing in Additive White Noise", IEEE Trans. Autom. Control, vol. AC-13, no. 6, pp. 655-660, Dec. 1968.

(208) Kale, B.K., "On the Solution of the Likelihood Equation by Iteration Processes", Biometrika, 48, pp. 452-456, 1961.

(209) Kalman, R.E., "A New Approach to Linear Filtering and Prediction Problems", Trans. ASME, vol. 82D, pp. 35, 1960.

(210) Kalman, R.E., Y.C. Ho and K.S. Nahendra, "Controllability of Linear Dynamical Systems", Contributions to Differential Equations, vol. 1, no. 2, pp. 189-213, 1962.

(211) Kalman, R.E., "On the General Theory of Control Systems, Proc. IFAC Moscow Congress, vol. 1, Butterworth, Inc., Washington, D.C., pp. 481-492, 1960.

(212) Kalman, R.E., "New Methods in Wiener Filtering Theory", Proc. First Symp. Eng. Appl. Random Functions Theory Probability, John Wiley & Sons, Inc., New York, 1963.

(213) Kalman, R.E., "When is a Linear Control System Optimal?" Trans. ASME, J. Basic Eng., vol. 86D, pp. 51-60, March 1964.

(214) Kalman, R.E., "Linear Stochastic Filtering Theory", Proc. Symp. System Theory, Polytechnic Press, Brooklyn, N.Y., pp. 197-206, 1965.

(215) Kalman, R.E., and J.E. Bertram, "Control System Analysis and Design via the "Second Method" of Liapunov, I, Continuous-time Systems", Trans. ASME, J. Basic Eng., vol. 82D, pp. 371-393, 1960.

(216) Kalman, R.E. and R. Bucy, "New Results in Linear Filtering and Prediction Theory", Trans. ASME, J. Basic Eng., vol. 83D, pp. 95-108, March 1961.

(217) Kalman, R.E., T.S. Englar and R.S. Bucy, "Fundamental Study of Adaptive Control Systems", Wright-Patterson Air Force Base, Ohio, ASD-TR 61-27, 1961.

(218) Kalman, R.E., P.L. Falb and M.A. Arbib, "Topics in Mathematical System Theory", McGraw-Hill Book Company, New York, 1969.

(219) Kaminski, P.G., A.E. Bryson and S.F. Schmidt, "Discrete Square Root Filtering: A Survey of Current Techniques", IEEE Trans. on Auto. Control, vol. AC-16, no. 6, pp. 727-736, 1971.

(220) Kasper, J.F., Jr., "A Skywave Correction Adjustment Procedure for Improved OMEGA Accuracy", Proceedings of the ION National Marine Meeting, New London, Connecticut, Oct. 1970.

(221) Kayton, M. and W. Fried, eds., Avionics Navigation Systems, John Wiley & Sons, Inc., New York, 1969.

(222) Kelly, E.J., I.S. Reed and W.L. Root, "The Detection of Radar Echoes in Noise. I", J. Soc. Indust. Appl. Math., 8, pp. 309-341, June 1960.

(223) Kelly, H.J., "Method of Gradients", chapter 6 of Optimization Techniques, edited by Leitmann, Academic Press, pp. 206-254.

(224) Kiefer, E. and J. Wolfowitz, "Stochastic Estimation of the Maximum of a Regression Function", Ann. Math. Statistics, vol. 23, no. 3, 1952.

(225) Klein, P.I., "Application of the Concept of Referenced Radio Navigation," IEEE Trans. Aerospace Electron. Systems, Vol. AES-4, no. 4, pp. 494-498, July 1968.

(226) Klementis, K.A. and C.J. Standish, "Final Report - Phase I - Synergistic Navigation System Study", IBM Corp., Owego, N.Y., IBM no. 67-923-7, (AD 678 070), October 1966.

(227) Klinger, A., "Prior Information and Bias in Sequential Estimation", IEEE Trans. Autom. Control, pp. 102-105, February 1968.

(228) Knoll, A. and M. Edelstein, "Estimation of Local Vertical and Orbital Parameters for an Earth Satellite Using Horizon Sensor Measurements", AIAA J., vol. 3, no. 2, pp. 338-345, 1965.

(229) Kolmogorov, A.N., "Interpolation and Extrapolation of Stationary Random Sequences", Trans. by W. Doyle and J. Selin, RM-3090-PR, The Rand Corp., Santa Monica, Calif., 1962.

(230) Kolmogorov, A.N., "Interpolation and Extrapolation von Stationaren Zufalligen Folgen", Bull. Acad. Sci. USSR, Ser. Math. 5, pp. 3-14, 1941.

(231) Korn, G.A. and T.M. Korn, Mathematical Handbook for Engineers and Scientists, McGraw-Hill Book Company, New York, 1961.

(232) Kozin, C.H. and G.R. Cooper, "The Use of Prior Estimates in Successive Point Estimations of a Random Signal", J. SIAM Appl. Math., vol. 14, no. 1, pp. 112-130, Jan. 1966.

(233) Kreindler, E. and P.E. Sarachik, "On the Concepts of Controllability and Observability of Linear Systems", IEEE Trans. Autom. Control, vol. AC-9, no. 2, pp. 129-136, April 1964.

(234) Kroy, W.H. and A.R. Stubberud, "Identification via Nonlinear Filtering," Intern. J. Control, vol. 6, no. 6, pp. 499-522, 1967.

(235) Kullback, S., Information Theory and Statistics, John Wiley & Sons, Inc., New York, 1959.

(236) Kuo, B.C., Analysis and Synthesis of Sampled-Data Control Systems, Prentice-Hall, Inc., Englewood Cliffs, N.J., 1963.

(237) Kushner, H.J., "On the Differential Equations Satisfied by Conditioned Probability Densities of Markov Processes, with Applications", J. SIAM Control, ser. A, vol. 2, no. 1, pp. 106-119, 1962.

(238) Kushner, H.J., "Dynamical Equations for Optimal Nonlinear Filtering", J. Differential Equations, vol. 3, pp. 179-190, 1967.

(239) Kushner, H.J., "Nonlinear Filtering: The Exact Dynamical Equations Satisfied by the Conditional Mode", IEEE Trans. Autom. Control, vol. AC-12, no. 3, pp. 262-267, June 1967.

(240) Kushner, H.J., "Approximations to Optimal Non-linear Filters", IEEE Trans. Autom. Control, vol. AC-12, no. 5, pp. 546-556, October 1967.

(241) Kushner, H.J., "On the Convergence of Lion's Identification Method with Random Inputs", IEEE Trans. on Automatic Control, Vol. AC-15, No. 6, December 1970.

(242) Kwakernaak, H., "Optimal Filtering in Linear Systems with Time Delays", IEEE Trans. Autom. Control, vol. AC-12, no. 2, pp. 169-173, April 1967.

(243) Laning, J.H., Jr. and R.H. Battin, Random Processes in Automatic Control, McGraw-Hill Book Company, New York, 1956.

(244) Larson, R.E. and J. Peschon, "A Dynamic Programming Approach to Trajectory Estimation", IEEE Trans. Autom. Control, vol. AC-11, no. 3, pp. 537-540, July 1966.

(245) Lasdon, L.S., S. Mitter and A. Warren, "The Method of Conjugate Gradient for Optimal Control Problems", IEEE Trans. Automatic Control, April 1967.

(246) Lawson, J.L. and G.E. Uhlenbeck, Threshold Signals, McGraw-Hill Book Company, New York, 1950.

(247) Leach, R., "Evaluating the Quality of Prediction for a Position-predicting or Tracking System", J. Amer. Rocket Soc., 32, pp. 1697-1701, November 1962.

(248) LeCam, L., "On the Asymptotic Theory of Estimation and Testing Hypotheses", Proceedings of the Third Berkeley Symposium on Mathematical Statistics and Probability, Berkeley and Los Angeles: University of California Press, pp. 129-156, 1956.

(249) Lee, B., "The Improvement on a Low-cost Inertial Navigator Through Conditional Feedback", IEEE Trans. Aerospace Electron. Systems, vol. AES-4, no. 4, pp. 627-634, July 1968.

(250) Lee, R.C.K. Optimal Estimation, Identification and Control, M.I.T. Res. Monograph 28, Cambridge, Mass., 1964.

(251) Lee, R.C.K., "A Moving Window Approach to the Problems of Estimation and Identification", Aerospace Corp., El Segundo, California, Report No. TR-1001 (2307)-23, June 1967.

(252) Lee, Y.W., Statistical Theory of Communication, John Wiley & Sons, Inc., New York, 1960.

(253) Legendre, A.M., Nouvelles Methodes Pour la Determination des Orbites des Cometes, Paris, 1806.

(254) Lehman, E.L., Testing Statistical Hypotheses, John Wiley & Sons, Inc., New York, 1959.

(255) Leondes, C.T., ed., Guidance and Control of Aerospace Vechicles, McGraw-Hill Book Co., Inc., New York, 1963.

(256) Leondes, C.T., Peller, J.B. and Stear, E.B., "Nonlinear Smoothing Theory", IEEE Trans. Systems Science and Cybernetics, Vol. SSC-6, No. 1, January 1970.

(257) Levenberg, K., "A Method for the Solution of Certain Non-linear Problems in Least Squares", Quart. Appl. Math., 2, pp. 164-168, 1944.

(258) Levine, N., "A New Technique for Increasing the Flexibility of Recursive Least Squares Data Smoothing", Bell System Tech. J., 40, pp. 821-840, May 1961.

(259) Liebelt, P.B., An Introduction to Optimal Estimation, Addison-Wesley Publishing Company, Inc., Reading, Mass., 1967.

(260) Light, D.L., "Photo Geodesy from Apollo", Photogrammetric Engineering, Vol. 38, No. 6, pp. 547-587, 1972.

(261) Lion, P.M., "Rapid Identification of Linear and Nonlinear Systems", AIAA Journal, Vol. 5, No. 10, October 1967.

(262) Loeve, M., Probability Theory, D. Van Nostrand Company, Inc. Princeton, N.J., 1955.

(263) Lucky, R.W., J. Salz, and E.J. Weldon, Jr., Principles of Data Communication, McGraw-Hill Book Company, New York, 1968.

(264) Luenberger, D.G., "Observing the State of a Linear System", IEEE Trans. Mil. Electron., vol. MIL-8, pp. 74-80, April 1964.

(265) Luenberger, D.G., "Observers for Multivariable Systems", IEEE Trans. Autom. Control, vol. AC-11, no. 2, pp. 190-197, April 1966.

(266) Magil, D.T., "Optimal Adaptive Estimation of Sampled Stochastic Processes", IEEE Trans. Autom. Control, vol. AC-10, no. 4, pp. 434-439, October 1965.

(267) Mahalanabis, A.K. and M. Farooq, "A Second-Order Method for State Estimation of Nonlinear Dynamical Systems", Int. J. Control, Vol. 14, No. 4, pp. 631-639, 1971.

(268) Marquardt, D.W., "An Algorithm for Least-Squares Estimation of Nonlinear Parameters", J. Soc. Indust. Appl. Math., 11, pp. 431-441, June 1963.

(269) Mayne, D.Q., "A Solution of the Smoothing Problem for Linear Dynamic Systems", Automatica, vol. 4, pp. 73-92, 1966.

(270) McAulary, R.J., "Numerical Optimization Techniques Applied to PPM Signal Design", IEEE Trans. Inform. Theory, vol. IT-14, no. 5, pp. 708-716, September 1968.

(271) McBride, A.L. and A.P. Sage, "Optimum Estimation of Bit Synchronization", IEEE Trans. Aerospace Electron. Systems, vol. 5, no. 3, pp. 525-536, May 1969.

(272) McBride, A.L. and A.P. Sage, "On the Discrete Sequential Estimation of Bit Synchronization", IEEE Trans. Comm. Technology, vol. 18, no. 1, pp. 48-58, February 1970.

(273) McGhee, R.B. and R.B. Walford, "A Monte Carlo Approach to the Evaluation of Conditional Expectation Parameter Estimates for Nonlinear Dynamic Systems", IEEE Trans. Autom. Control, vol. AC-13, no. 1, pp. 29-36, February 1968.

(274) McLendon, J.R. and A.P. Sage, "Computational Algorithms for Discrete Detection and Likelihood Ratio Computation", Information Sciences, vol. 2, no. 3., 1970.

(275) McReynolds, S.R., "A New Approach to Stochastic Calculus", presented at the Seminar on Guidance Theory and Trajectory Analysis, May 31, 1967, NASA Electronics Research Center, Cambridge, Mass.

(276) Meditch, J.S., "Orthogonal Projection and Discrete Optimal Linear Smoothing", SIAM J. Control, vol. 5, no. 1, pp. 74-80, February 1967.

(277) Meditch, J.S., "Optimal Fixed Point Continuous Linear Smoothing", Proc. Joint Autom. Control Conf., pp. 249-257, June 1967.

(278) Meditch, J.S., "The Wiener-Hopf Solution and the Optimal Fixed Point Smoothing Problem", Inform-processing Control Systems Lab. Tech. Rept., 67-111, June 1967.

(279) Meditch, J.S., "On Optimal Fixed Point Linear Smoothing", Boeing Sci. Res. Lab. Inform. Sci. Rept. 1, August 1967.

(280) Meditch, J.S., "On Optimal Linear Smoothing Theory", Inform. Control, vol. 10, pp. 598-615, 1967.

(281) Meditch, J.S., "A Successive Approximation Procedure for Nonlinear Data Smoothing", Proc. Symp. Inform. Processing, April 1969, pp. 555-568, Purdue Univ., Lafayette, Ind.

(282) Meditch, J.S., "Stochastic Optimal Linear Estimation and Control", McGraw-Hill Book Company, New York, 1969.

(283) Mehra, R.K., "On Optimal and Suboptimal Linear Smoothing", Proc. 1968 Nat. Electron. Conf., pp. 119-124, 1968.

(284) Mehra, R.K., and A.E. Bryson, Jr., "Linear Smoothing Using Measurements Containing Correlated Noise with an Application to Inertial Navigation," IEEE Trans. Autom. Control, vol. AC-13, no. 5, pp. 496-503, October 1968.

(285) Mehra, R.K. and A.E. Bryson, Jr., "Smoothing for Time-varying Systems Using Measurements Containing Colored Noise", Proc. Joint Autom. Control Conf., Ann Arbor, Mich., pp. 871-883, 1968.

(286) Meirovitch, L., Methods of Analytical Dynamics, McGraw-Hill Book Company, 1970.

(287) Melsa, J.L., "Frequency Domain Derivation of the Stationary Kalman Filter", Proc. First Ann. Houston Conf. Circuits Systems Computers, pp. 323-332, 1969.

(288) Melsa, J.L. and D. Schultz, Linear Control Systems, McGraw-Hill Book Company, New York, 1969.

(289) Merson, R.H., "Numerical Integration of Differential Equations ...", Contract Report ESOC 377/71/HR, published by the Royal Aircraft Establishment, Farnborough, Hants, Great Britain, May 1973.

(290) Mikhail, E.M., Observations and Least Squares, IEP-A Dun-Donnelly Publishers, New York, 1977.

(291) Middleton, D., Introduction to Statistical Communication Theory, McGraw-Hill Book Company, New York, 1960.

(292) Middleton, D. and R. Esposito, "Simultaneous Optimum Detection and Estimation of Signals in Noise", IEEE Trans. Inform. Theory, vol. IT-14, no. 3, pp. 434-444, May 1968.

(293) Middleton, D. and D. Van Meter, "Detection and Extraction of Signals in Noise from the Point of View of Statistical Decision Theory", J. Soc. Ind. Appl. Math., vol. 3, p. 192, 1955; vol. 4, p. 86, 1956.

(294) Miller, R.W., "Asymptotic Behavior of the Kalman Filter with Exponential Aging", AIAA Journal, vol. 9, no. 3, pp. 537-539, March 1971.

(295) Milne, W.E., Numerical Solution of Differential Equations, Dover Publishing Company, New York, 1970.

(296) Monroe, A.J., Digital Processes for Sampled Data Systems, New York, John Wiley and Sons, Inc., 1962.

(297) Mood, A.M. and F.A. Graybill, Introduction to the Theory of Statistics, 2nd ed., McGraw-Hill Book Company, New York, 1963.

(298) Mood, A.M., Introduction to the Theory of Statistics, New York, McGraw-Hill Book Company, 1950.

(299) Moore, J.B., "Optimum Differentiation Using Kalman Filter Theory", Proc. IEEE, vol. 56, no. 5, p. 871, May 1968.

(300) Morrison, D., "A Method for Nonlinear Minimization Problems", Internal Report No. NN-140, Space Technology Labs, Redondo Beach, Calif., 1959.

(301) Murphy, G.J., and K. Sahara, "A Mean-weighted Square-error Criterion for Optimum Filtering of Nonstationary Random Processes," IRE, Trans. PGAC, AC-6, pp. 211-216, May 1961.

(302) Nahi, N.E. and B.M. Schaefer, "Decision-Directed Adaptive Recursive Estimators: Divergence Prevention", IEEE Trans. on Auto. Control, vol. AC-17, no. 1, pp. 61-67, Feb. 1972.

(303) Nash, R.A., Jr., J.A. D'Appolito, and K.J. Roy, "Error Analysis of Hybrid Inertial Navigation Systems", AIAA Guidance and Control Conference, Stanford, Calif., Aug. 1972.

(304) Nash, R.A., Jr., J.F. Kasper, Jr., B.S. Crawford and S.A. Levine, "Application of Optimal Smoothing to the Testing and Evaluation of Inertial Navigation Systems and Components", IEEE Trans. on Automatic Control, vol. AC-16, No. 6, December 1971.

(305) Nash, R.A., Jr., S.A. Levine and K.J. Roy, "Error Analysis of Space-Stable Inertial Navigation Systems", IEEE Trans. on Aero. and Electronic Systems, vol. AES-7, no. 4, pp. 617-629, July 1971.

(306) Nash, R.A., Jr., "The Estimation and Control of Terrestrial Inertial Navigation System Errors Due to Vertical Deflections", IEEE Trans. Autom. Control, vol. AC-13, no. 4, pp. 329-338, August 1968.

(307) Nash, R.A., Jr. and F.B. Tuteur, "The Effect of Uncertainties in the Noise Covariance Matrices on the Maximum Likelihood Estimate of a Vector", IEEE Trans. Autom. Control, vol. 13, no. 1, pp. 86-88, February 1968.

(308) Nayfeh, A., Perturbation Methods, Wiley Publishing Company, New York, 1973.

(309) Neal, S.R., "Linear Estimation in the Presence of Errors in Assumed Plant Dynamics", IEEE Trans. Autom. Control, vol. AC-12, no. 5, pp. 592-594, October 1967.

(310) Newbold, P.M. and Y.C. Ho, "Detection of Changes in Characteristics of a Gauss-Markov Process", IEEE Trans. Aerospace Electron. Systems, vol. AES-4, no. 5, pp. 707-718, September 1968.

(311) Newton, G.C., et al, Analytical Design of Linear Feedback Controls, John Wiley & Sons, Inc., New York, 1957.

(312) Neyman, J. and E.S. Pearson, "The Testing of Statistical Hypotheses in Relation to Probability a Priori", Proc. Cambridge Phil. Soc., 29, 1933.

(313) Neyman, J. and E.S. Pearson, "Contributions to the Theory of Testing Statistical Hypotheses", Stat. Res. Mem., Pts. I and II, 1936, 1938.

(314) Neyman, J., "Outline of a Theory of Statistical Estimation Based on the Classical Theory of Probability", Philos. Trans. Royal Soc. London, Ser. A., 236, pp. 333-380, 1937.

(315) Neyman, J., "Contributions to the Theory of the χ^2 Test", Proceedings of the Berkeley Symposium on Mathematical Statistics and Probability, Berkeley and Los Angeles: Univ. of Calif. Press, pp. 239-273, 1949.

(316) Nicoliades, J.D., Lectures on Free Flight Missile Dynamics, Dept. of Aerospace Engineering, Univ. of Notre Dame, 1967.

(317) Nishimura, T., "On the A-priori Information in Sequential Estimation Problems", IEEE Trans. Autom. Control, vol. AC-11, no. 2, pp. 197-204, April 1966.

(318) Nishimura, T., "Correction to and Extension of "On the A-priori Information in Sequential Estimation Problems", IEEE Trans. Autom. Control, vol. AC-12, p. 123, February 1967.

(319) Nishimura, T., "Error Bounds of Continuous Kalman Filters and the Application to Orbit Determination Problems", IEEE Trans. Autom. Control, vol. AC-12, no. 3, pp. 268-275, June 1967.

(320) North, D.O., "Analysis of the Factors Which Determine Signal/ Noise Discrimination in Pulse Carrier Systems", Proc. IRE, vol. 51, pp. 1016-1028, July 1963.

(321) Novak, L.M., "Optimal Minimal-Order Observers for Discrete-Time Systems - A Unified Theory", Automatica, vol. 8, July 1972.

(322) Ogata, K., State Space Analysis for Engineers, Prentice Hall, pp. 120-173, 1967.

(323) O'Halloran, W.F., Jr. and B.J. Uttam, "Derivation of Observers for Continuous Linear Stochastic Systems From the Discrete Formulation", Sixth Asilomar Conference, Pacific Grove, California, November 1972.

(324) O'Halloran, W.F., Jr., "A Suboptimal Error Reduction Scheme for a Long-Term Self-Contained Inertial Navigation System", National Aerospace and Electronics Conference, Dayton, Ohio, May 1972.

(325) Ohap, R.F. and A.R. Stubberud, "A Technique for Estimating the State of a Nonlinear System", IEEE Trans. Autom. Control, vol. AC-10, no. 2, pp. 150-155, April 1965.

(326) Omura, J.K., "Signal Optimization for Additive Noise Channels with Feedback", presented at WESCON, Session 7, The Application of State-variable Techniques to Communication and Radar Problems, August 1966.

(327) O'Neill, E.L., "Spatial Filtering in Optics", IRE, Trans. PGIT, IT-2, pp. 56-65, June 1956.

(328) Oppenheim, A.V., R.W. Schafer and T.G. Stockham, Jr., "Non-linear Filtering of Multiplied and Convolved Signals", Proc. IEEE, vol. 56, no. 8, pp. 1264-1291, August 1968.

(329) Osborne, E.E., "On Least Squares Solutions of Linear Equations", J. Assoc. Comput. Mach., 8, pp. 628-636, 1961.

(330) Ostrowski, A.M., "Solution of Equations and Systems of Equations," 2nd ed., Academic Press, New York,

(331) Papoulis, A., Probability, Random Variables and Stochastic Processes, McGraw-Hill Book Company, New York, 1965.

(332) Parzen, E., Modern Probability Theory and Its Applications, New York, Wiley Publishing Company, 1960.

(333) Parzen, E., Stochastic Processes, San Francisco, Calif, Holden-Day, 1962.

(334) Pearson, J.B., "On Nonlinear Least-squares Filtering", Automatica, vol. 4, pp. 97-105, 1967.

(335) Penrose, R., "A Generalized Inverse for Matrices", Proc. Cambridge Phil. Soc., 51, pp. 406-413.

(336) Penrose, R., "On Best Approximate Solution of Linear Matrix Equations", Proc. Cambridge Philos. Soc. 52, pp. 17-19, 1956.

(337) Pentecost, E.E. and A.R. Stubberud, "Synthesis of Computationally Efficient Sequential Linear Estimators", IEEE Trans. Aerospace Electron. Systems, vol. AES-3, no. 2, pp. 242-249, March 1967.

(338) Peters, G. and J.H. Wilkinson, "The Least Squares Problem and Pseudoinverses", Comput. J., 13, pp. 309-316, 1970.

(339) Peterson, W.W., T.G. Birdsall and W.C. Fox, "The Theory of Signal Detectability", IRE Trans. Inform. Theory, PGIT-4, p. 171, September, 1954.

(340) Phaneuf, R.J., Approximate Nonlinear Estimation, Ph.D. Thesis, Massachusetts Institute of Technology, May 1968.

(341) Pickholtz, R.L. and R.R. Boorstyn, "A Recursive Approach to Signal Detection", IEEE Trans. Inform. Theory, vol. IT-14, no. 3, pp. 445-450, May 1968.

(342) Potter, J.E. and D.C. Fraser, "A Formula for Updating the Determinant of the Covariance Matrix", AIAA J., vol. 5, no. 7, pp. 1352-1354, July 1967.

(343) Potter, J.E., "A Guidance-Navigation Separation Theorem", Report RE-11, Experimental Astronomy Laboratory, M.I.T., August 1964.

(344) Price, C.F., "An Analysis of the Divergence Problem in the Kalman Filter", IEEE Trans. on Auto. Control, Vol. AC-13, no. 6, December 1968.

(345) Price, C.F., R.S. Warren, A. Gelb, and W.E. Vander Velde, "Evaluation of Homing Guidance Laws Using the Covariance Analysis Describing Function Technique", Trans. First NWC Symposium on the Application of Control Theory to Modern Weapons Systems", China Lake, pp. 73-94, June 1973.

(346) Pugachev, V.S., "The Method of Determining Optimum Systems Using Bayes Criteria", IRE Trans. PGCT, CT-7, pp. 491-505, December 1960.

(347) Pugachev, V.S., Theory of Random Functions, Pergamon Press, New York, 1963.

(348) Raiffa, H. and R. Schlaifer, Applied Statistical Decision Theory, Cambridge, Mass, Harvard University Press, 1961.

(349) Rao, C.R., "Information and Accuracy Attainable in the Estimation of Statistical Parameters", Bull. Calcutta Math. Soc., vol. 37, pp. 81-91, 1945.

(350) Rao, C.R., Linear Statistical Inference and Applications, John Wiley & Sons, Inc., New York, 1965.

(351) Rao, C.R., "Analysis of Dispersion for Multiplying Classified Data with Unequal Numbers in Cells", Sankhya, pp. 253-280, 1955.

(352) Rao, C.R., "A Note on a Generalized Inverse of a Matrix with Applications to Problems in Mathematical Statistics," J. Roy. Statis. Soc., Ser. B, 24, pp. 152-158, 1962.

(353) Rao, C.R. and S.K. Mitra, Generalized Inverse of Matrices and Its Applications, Wiley, New York, 1971.

(354) Rauch, H.E.,"Solutions to the Linear Smoothing Problem", IEEE Trans. Autom. Control, vol. AC-8, no. 4, pp. 371-372, October 1963.

(355) Rauch, H.E., "Optimum Estimation of Satellite Trajectories Including Random Fluctuations in Drag", AIAA J., vol. 3, no. 4, pp. 717-722, 1965.

(356) Rauch, H.E., "Least Squares Estimation with a Large Number of Parameters", Proc. Joint Autom. Control Conf., pp. 241-248, 1967.

(357) Rauch, H.E., F. Tung and C.T. Striebel, "Maximum Likelihood Estimates of Linear Dynamic Systems", AIAA J., vol. 3, no. 8, pp. 1445-1450, August 1965.

(358) Reid, J.K., ed., Large Sparse Sets of Linear Equations, Academic Press, New York, 1971.

(359) Rice, S.O., "Mathematical Analysis of Random Noise", Bell System Tech. J., vol. 23, pp. 282-332, 1944.

(360) Richards, F.S.G., "A Method of Maximum-Liklihood Estimation", J. Royal Statistical Soc., Ser. B., 23, pp. 469-475, 1961.

(361) Richman, J. and B. Friedland, "Design of Optimum Mixer-Filter for Aircraft Navigation Systems", Natl. Aerospace Electron. Conf. Proc., Dayton, Ohio, pp. 429-438, May 1967.

(362) Riddle, A.C. and B.D.O. Anderson, "Spectral Factorization-Computational Aspects", IEEE Trans. Autom. Control, vol. AC-11, no. 4, pp. 764-765, October 1966.

(363) Roberts, A.P., "Optimal Linear Filtering and Lagging Filtering of Coloured Noise", Intern. J. Control, vol. 8, no. 4, pp. 401-416, 1968.

(364) Robbins, H. and S. Munroe, "A Stochastic Approximation Method", Ann. Math. Statistics, vol. 22, no. 1, pp. 400-407, 1951.

(365) Rose, D.J. and R.A. Willoughby, Sparse Matrices and Their Applications, Plenum Press, New York, 1972.

(366) Rose, R.E. and G.M. Lance, "Analysis of the Behavior of Continuous Parameter Estimation Methods", Proc. Joint Autom. Control Conf., Ann Arbor, Mich., pp. 109-115, June 1968.

(367) Rosenblatt, M., Random Processes, Oxford University Press, New York, 1962.

(368) Sacks, J.E. and H.W. Sorenson, "Comment on 'A Practical Nondiverging Filter'", AIAA Journal, Vol. 9, No. 4, pp. 767, 768, April 1971.

(369) Sage, A.P. and J.L. Melsa, Estimation Theory, McGraw-Hill, pp. 175-237, 1971.

(370) Sage, A.P., Optimum Systems Control, Prentice-Hall, Inc., Englewood Cliffs, N.J., 1968.

(371) Sage, A.P., "Quasilinear Techniques for Filtering in Discrete Nonlinear Systems", Proc. Third Ann. Princeton Conf. Inform. Sci. Systems, pp. 452-456, March 1969.

(372) Sage, A.P., "Maximum A-posteriori Filtering and Smoothing Algorithms", Intern. J. Control, vol. 11, no. 3, May 1970.

(373) Sage, A.P. and W.S. Ewing, "On Smoothing Algorithms for Nonlinear State and Parameter Estimation", Proc. Second Hawaii Intern. Conf. System Sci., pp. 373-376, January 1969.

(374) Sage, A.P. and W.S. Ewing, "On Filtering and Smoothing Algorithms for Nonlinear State Estimation," Intern J. Control, vol. 11, no. 1, pp. 1-18, January 1970.

(375) Sage, A.P. and G.W. Husa, "Adaptive Filtering with Unknown Prior Statistics", Proc. 1969 Joint Autom. Control Conf., pp. 760-769, August 1969.

(376) Sage, A.P. and J.R. McLendon, "Discrete Sequential Detection and Likelihood Ratio Computation for Non-Gaussian Signals in Gaussian Noise", Proc. Symp. Inform. Processing, pp. 589-598, Purdue University, Lafayette, Ind., April 1969.

(377) Sage, A.P. and G.W. Masters, "On-line Estimation of States and Parameters for Discrete Nonlinear Dynamic Systems", Proc. Natl. Electron. Conf., pp. 677-682, 1966.

(378) Sage, A.P. and G.W. Masters, "Identification and Modeling of States and Parameters of Nuclear Reactor Systems", IEEE Trans. Nuclear Systems, vol. NS-14, no. 1, pp. 279-285, February 1967.

(379) Sage, A.P. and G.W. Masters, "Least Squares Curve Fitting and Discrete Optimal Filtering", IEEE Trans. Educ., vol. E-10, no. 1, pp. 29-36, March 1967.

(380) Sage, A.P. and J.L. Melsa, System Identification, Academic Press Inc., New York, 1970.

(381) Sain, M.K., "On the Control Applications of a Determinant Equality Related to Eigenvalue Computation", IEEE Trans. Autom. Control, vol. AC-11, no. 1, pp. 109-111, January 1966.

(382) Sakrison, D., "The Use of Stochastic Approximation to Solve the System Identification Problem", IEEE Trans. Autom. Control, vol. AC-12, no. 5, pp. 563-567, October 1967.

(383) Sakrison, D., Communication Theory: Transmission of Wave Forms and Digital Information, John Wiley & Sons, Inc., New York, 1968.

(384) Saridis, G.N., "Learning Applied to Successive Approximation Algorithms", Proc. Joint Autom. Control Conf., pp. 1007-1013, Ann Arbor, Mich., June 1968.

(385) Saridis, G.N. and G. Stein, "Stochastic Approximation Algorithms for Linear Discrete Time System Identification", IEEE Trans. Autom. Control, vol. AC-13, no. 5, pp. 515-523, October 1968.

(386) Sarma, V.V.S. and B.L. Deekshatulu, "Optimal Control When Some of the State Variables Are Not Measurable", Intern. J. Control, vol. 7, no. 3, pp. 251-156, 1968.

(387) Sawaragi, Y., Y.Sunahara and T. Nakamizo, Statistical Decision Theory in Adaptive Control Systems, Academic Press Inc., New York, 1967.

(388) Scheffe, H., The Analysis of Variance, New York, John Wiley and Sons, Inc., 1959.

(389) Schlee, F.H., C.J. Standish and N.F. Toda, "Divergence in the Kalman Filter," AIAA J., vol. 5, no. 6, pp. 1114-1120, June 1967.

(390) Schlegel, L.B., "Covariance Matrix Approximation", AIAA J., vol. 1, pp. 2672-2673, November 1963.

(391) Schlesinger, F., "On the Errors in the Sum of a Number of Tabular Quantities", Astronom. J., 30, pp. 183-190, 1917.

(392) Schmetterer, L., "Stochastic Approximation", Proc. 4th Berkeley Symp. on Mathematical Statistics and Probability, Vol. 1, Los Angeles Univ. of Calif. Press, pp. 587-609, 1958.

(393) Schmidt, S.F., "Computational Techniques in Kalman Filtering", Theory and Applications of Kalman Filtering, Advisory Group for Aerospace Research and Development, AGARDograph 139, (AD 704 306), February 1970.

(394) Schmidt, S.F., et al, "Case Study of Kalman Filtering in the C-5 Aircraft Navigation System", Case Studies in System Control sponsored by IEEE Group on Autom. Control, 1968 Joint Autom. Control Conf., Ann Arbor, Mich., 1968.

(395) Schultz, D.G. and J.L. Melsa, State Functions and Linear Control Systems, McGraw-Hill Book Company, New York, 1967.

(396) Schutz, B.E., "Analytical Approach to Orbit Determination in the Presence of Model Errors", Amer. Inst. of Aeronautics and Astronautics paper #73-170, presented to the 11th AIAA Aerospace Sciences Meeting, Wash, D.C., 1973.

(397) Schwartz, L. and R. W. Bass, "Extensions to Optimal Multichannel Nonlinear Filtering", Hughes Aircraft Co., Space Sys. Div., Rept. SSD 60220R, February 21, 1966.

(398) Schwartz, L. and E.B. Stear, "A Valid Mathematical Model for Approximate Nonlinear Minimal-variance Filtering", J. Math. Anal. Appl., vol. 21, pp. 1-6, January 1968.

(399) Schwartz, L. And E.B. Stear, "A Computational Comparison of Several Nonlinear Filters", IEEE Trans. Autom. Control, vol. 13, no. 1, pp. 83-86, February 1968.

(400) Schwartz, M., "Abstract Vector Spaces Applied to Problems in Detection and Estimation Theory", IEEE Trans. Inform. Theory, vol. IT-12, no. 3, pp. 327-336, July 1966.

(401) Schwartz, M., W.R. Bennett, and S. Stein, Communication Systems and Techniques, McGraw-Hill Book Company, New York, 1966.

(402) Schweppe, F.C., "Optimization of Signals", M.I.T. Group Rept. 1964-4, Jan. 16, 1964.

(403) Schweppe, F.C., "Evaluation of Likelihood Functions for Gaussian Signals", IEEE Trans. Inform. Theory, vol. IT-11, no. 1, pp. 61-70, January 1965.

(404) Schweppe, F.C., "Radar Frequency Modulations for Accelerating Targets Under a Bandwidth Constraint", IEEE Trans. Mil. Electron., vol. MIL-9, no. 1, pp. 25-32, January 1965.

(405) Schweppe, F.C., "Recursive State Estimation Unknown but Bounded Errors and System Inputs", IEEE Trans. Autom. Control, vol. AC-13, no. 1, pp. 22-28, February 1968.

(406) Schweppe, F.C., "Sensor-array Data Processing for Multiple-signal Sources, IEEE Trans. Inform. Theory, vol. IT-14, no. 2, pp. 294-305, March 1968.

(407) Schweppe, F.C. and D.L. Gray, "Radar Signal Design Subject to Simultaneous Peak and Average Power Constraints", IEEE Trans. Inform. Theory, vol. IT-12, no. 1, pp. 13-26, January 1966.

(408) Schweppe, F.C., Uncertain Dynamic Systems, Prentice-Hall, Englewood Cliffs, N.J., 1973.

(409) Seidman, L.P., "An Upper Bound on Average Estimation Error in Nonlinear Systems", IEEE Trans. Inform. Theory, vol. IT-14, no. 2, pp. 243-249, March 1968.

(410) Selin, I., "The Sequential Estimation and Detection of Signals in Normal Noise, I, Inform. Control, vol. 7, pp. 512-534, 1964.

(411) Selin, I., "The Sequential Estimation and Detection of Signals in Normal Noise, II", Inform. Control, vol. 8, pp.1-35, 1965.

(412) Selin, I., Detection Theory, Princeton University Press, Princeton, New Jersey, 1965.

(413) Shenton, L.R., "Maximum Likelihood and the Efficiency of the Method of Moments", Biometrika, 37, pp. 111-116, 1950.

(414) Shenton, L.R., "Moment Estimators and Maximum Likelihood", Biometrika, 45, pp. 311-320, 1958.

(415) Shenton, L.R., "The Distribution of Moment Estimators", Biometrika, 46, pp. 296-305, 1959.

(416) Sherman, S., "A Theorem on Convex Sets with Applications", Ann. Math. Statist., 26, pp. 763-767, 1955.

(417) Sherman, S., "Non-mean-square Error Criteria", IRE, Trans. PGIT, IT-4, pp. 125-126, Sept. 1958.

(418) Shinbrot, M., "A Generalization of a Method for the Solution of the Integral Equation Arising in Optimization of Time-Varying Linear Systems with Nonstationary Inputs", IRE, Trans. PGIT, IT-3, pp. 220-224, December 1957.

(419) Shinbrot, M., "Optimization of Time-Varying Linear Systems with Nonstationary Inputs", Trans. ASME, Vol. 80, pp. 457-462, 1958.

(420) Shutz, B.E., J.D. McMillan and B.D. Tapley, "A Comparison of Estimation Methods for the Reduction of Laser Observations of a Near-Earth Satellite", presented to the AAS/AIAA Astrodynamics Conference, Vail, Colorado, July 1973.

(421) Silvey, S.D., "A Note on the Maximum-likelihood in the Case of Dependent Random Variables", J. Roy. Statist. Soc., Ser. B, 23, pp. 444-452, 1961.

(422) Sims, C.S. and J.L. Melsa, "Specific Optimal Estimation", IEEE Trans. Autom. Control, vol. AC-14, no. 2, pp. 183-186, 1969.

(423) Singer, R.A. and P.A. Frost, "On the Relative Performance of the Kalman and Wiener Filters", IEEE Trans. on Auto. Control, vol. AC-14, no. 4, pp. 391-394, August 1969.

(424) Singer, R.A., "Estimating Optimal Tracking Filter Performance for Manned Maneuvering Targets", IEEE Trans. Aerosp. Elect. Syst., AES-6, pp. 473-483, July 1970.

(425) Sinha, N.K. and M.P. Griscik, "A Stochastic Approximation Method", IEEE Trans. on Systems, Man, and Cybernetics, Vol. SMC-1, no. 4, pp. 338-343, October 1971.

(426) Skolnik, M.I., Introduction to Radar Systems, New York, McGraw-Hill Book Company, 1962.

(427) Skorokhod, A.V., Studies in the Theory of Random Processes, Addison-Wesley Publishing Company, Inc., Reading, Mass., 1965.

(428) Slepian, D., "Estimation of Signal Parameters in the Presence of Noise", IRE, Trans. PGIT, IT-3, pp. 68-89, March 1964.

(429) Smith, G.L., "The Scientific Inferential Relationships Between Statistical Estimation, Decision Theory, and Modern Filter Theory", Proc. JACC, Rensselaer Polytechnic Inst., pp. 350-359, June 1965.

(430) Smith, G.L., "On the Theory and Methods of Statistical Inference", NASA Tech. Rept. NASA TR R-251, April 1967.

(431) Smith, G.L., "Sequential Estimation of Observation Error Variances in a Trajectory Estimation Problem", AIAA J., vol. 5, no. 11, pp. 1964-1970, November 1967.

(432) Smith, G.L. and S.F. Schmidt, "Application of Statistical Filter Theory to the Optimal Estimation of Position and Velocity on Board a Circumlunar Vehicle", NASA Rept. NASA-TR R-135, 1962.

(433) Snyder, D.L., "The State-variable Approach to Analog Communication Theory", IEEE Trans. Inform. Theory, vol. IT-14, no. 1, pp. 94-104, January 1968.

(434) Snyder, D.L., The State-Variable Approach to Continuous Estimation with Applications to Analog Communication Theory, M.I.T. Res. Monograph 51, Cambridge, Mass., 1969.

(435) Solloway, C.B., "Elements of Orbit Determination", Jet Propulsion Laboratory, Engineering Planning Document Report #255, Pasadena, Ca., 1964.

(436) Solodovnikov, V.V., Introduction to the Statistical Dynamics of Automatic Control Systems, Dover Publications, Inc., New York, 1960.

(437) Soong, T.T., "On a Priori Statistics in Minimum-variance Estimation Problems", J. Basic Eng., pp. 109-112, March 1965.

(438) Sorenson, H.W., "Kalman Filtering Techniques", Advan. Control Systems, vol. 3, pp. 219-292, 1966.

(439) Sorenson, H.W., "On the Error Behavior in Linear Minimum Variance Estimation Problems", IEEE Trans. Autom. Control, vol. AC-12, no. 5, pp. 557-562, October 1967.

(440) Sorenson, H.W., "Least-Squares Estimation: From Gauss to Kalman", IEEE Spectrum, pp. 63-68, July 1970.

(441) Stear, E.B. and A.R. Stubberud, "Optimal Filtering for Gauss-Markov Noise", Intern. J. Control, vol. 8, no. 2, pp. 123-130, 1264-1291, 1968.

(442) Stein, S. and J.J. Jones, Modern Communication Principles with Application to Digital Signaling, McGraw-Hill Book Company, New York, 1967.

(443) Steinway, W.J. and J.L. Melsa, "Discrete Estimation Algorithm for Randomly Sampled Stochastic Signals", IEEE Autom. Control, vol. AC-15, no. 3, June 1970.

(444) Stewart, G.W., Introduction to Matrix Computations, Academic Press, New York, N.Y., ch. 1, 3, 5, 1973.

(445) Stoep, D.R.V., "Trajectory Shaping for the Minimization of State Variable Estimation Errors", Proc. Joint Autom. Control Conf., pp. 884-891, Ann Arbor, Mich., June 1968.

(446) Stoer, J., "On the Numerical Solution of Constrained Least-Squares Problems", SIAM J. Numer. Anal. 8, 382-411, 1971.

(447) Stratonovich, R.L., Topics in the Theory of Random Noise, vols. 1 and 2, Gordon and Breach, Science Publishers, Inc., New York, 1963.

(448) Stratonovich, R.L., Conditional Markov Processes and Their Application to the Theory of Optimal Control, transl. by R.N. and N.B. McDonough, American Elsevier Publishing Company, Inc., New York, 1968.

(449) Stubberud, A.R., "Optimal Filtering for Gauss-Markov Noise", Aerospace Rept. No. TR-0158 (3307-01)-10, December 1967.

(450) Sunahara, Y., "An Approximate Method of State Estimation for Nonlinear Dynamical Systems", Joint Automatic Control Conference, Univ. of Colorado, 1969.

(451) Sutherland, A.A., Jr. and Gelb, A., "Application of the Kalman Filter to Aided Inertial Systems", Naval Weapons Center, China Lake, California, NWC TP 4651, August 1968.

(452) Swerling, P., "First Order Error Propagation in a Stagewise Smoothing Procedure for Satellite Observations", J. Astronautical Sciences, vol. 6, pp. 46-52, 1959.

(453) Swerling, P., "Note on a New Computational Data Smoothing Procedure Suggested by Minimum Mean Square Error Estimation", IEEE Trans. Inform. Theory, vol. 12, no. 1, pp. 9-12, January 1966.

(454) Tarn, T.S. and J. Zaborsky, "A Practical Nondiverging Filter", AIAA J., vol. 8, no. 6, pp. 1127-1133, June 1970.

(455) Thompson, J.S. and E.L. Titlebaum, "The Design of Optimal Radar Waveforms for Clutter Rejection Using the Maximum Principle", Suppl. IEEE Trans. Aerospace Electron. Sys., vol. AES-3, no. 6, pp. 581-589, November 1967.

(456) Thompson, M.M., Manual of Photogrammetry, American Society of Photogrammetry, 3rd ed., vol. 1, p. 469.

(457) Thornton, C.L. and G.J. Bierman, "A Numerical Comparison of Discrete Kalman Filtering Algorithms: An Orbit Determination Case Study", JPL Technical Memorandum 33-771, to appear in Automatica, Jan. 1977.

(458) Tinkleman, M. and J.D. Pearson, "A Decomposition Approach to the Optimal Smoothing Problem", JACC Preprints, pp. 646-654, 1967.

(459) Titlebaum, E.L., "Optimization Methods for Signal Design and Processing", Suppl. IEEE Trans. Aerospace Electron. Sys., vol. AES-3, no. 6, pp. 543-551, 1967.

(460) Toda, N.F., F.H. Schlee and P. Obsharsky, "The Region of Kálman Filter Convergence for Several Autonomous Navigation Modes", AIAA J., vol. 7, no. 4, pp. 622-627, 1969.

(461) Truxal, J.G., Automatic Feedback Control System Synthesis, McGraw-Hill Book Co., Inc., New York, 1955.

(462) Tse, E. and M. Athans, "Optimal Minimal-Order Observer Estimators for Discrete Linear Time-Varying Systems", IEEE Transactions on Automatic Control, vol. AC-15, no. 4, August 1970.

(463) Tufts, D.W. and D.A. Shnidman, "Optimum Waveforms Subject to Both Energy and Peak-value Constraints", Proc. IEEE, vol. 52, no. 9, pp. 1002-1007, September 1964.

(464) Tung, F., "Linear Control Theory Applied to Interplanetary Guidance", IEEE Trans. Autom. Control, vol. AC-9, no. 1, pp. 82-89, January 1964.

(465) Turin, G.L., "An Introduction to Matched Filters", IRE Trans. Inform. Theory, vol. IT-6, pp. 311-329, June 1960.

(466) VanTrees, H.L., Detection, Estimation and Modulation Theory, Part I, John Wiley & Sons, Inc., New York, 1968.

(467) Viterbi, A.J., Principles of Coherent Communication, McGraw-Hill Book Company, New York, 1966.

(468) Wainstein, L.A. and V.D. Zubakov, Extraction of Signals from Noise, trans. from the Russian by R.A. Silverman, Englewood, N.J., Prentice-Hall, Inc., 1962.

(469) Wald, A., Sequential Analysis, John Wiley & Sons, Inc., New York, 1947.

(470) Wald, A., Statistical Decision Theory, John Wiley & Sons, Inc., New York, 1949.

(471) Wax, N., ed., Collected Papers on Noise and Stochastic Processes, New York, Dover, 1954.

(472) Weiss, I.M., "A Survey of Discrete Kalman-Bucy Filtering with Unknown Noise Covariances", AIAA Guidance, Control and Flight Mechanics Conference, Paper No. 70-955, Santa Barbara, Calif., August 1970.

(473) Whittle, P., Prediction and Regulation by Linear Least-Square Methods, London, The English Universities Press, Ltd., 1963.

(474) Wiener, N., "Generalized Harmonic Analysis", Acta Math., vol. 55, p. 117, 1930.

(475) Wiener, N., The Extrapolation, Interpolation, and Smoothing of Stationary Time Series with Engineering Applications, John Wiley & Sons, Inc., New York, 1949.

(476) Wilkinson, J.H., "Error Analysis of Transformations Based on the Use of Matrices of the Form $I-2ww^H$", In Error in Digital Computation, L.B. Rall, ed., vol. 2, pp. 77-101, Wiley, New York.

(477) Wilks, S.S., Mathematical Statistics, John Wiley & Sons, Inc., New York, 1962.

(478) Wishner, R.P., J.A. Tabaczynski, and M. Athans, "A Comparison of Three Nonlinear Filters", Automatica, vol. 5, no. 4, pp. 487-496, July 1969.

(479) Wolfowitz, J., "On the Stochastic Approximation Method of Robbins and Munroe", Ann. Math. Statistics, vol. 23, no. 3, pp. 457-466, 1952.

(480) Wolovich, W.A., "On State Estimation of Observable Systems", Proc. Joint Autom. Control Conf., pp. 210-220, Ann Arbor, Mich., June 1968.

(481) Wong, E. and J.B. Thomas, "On the Multidimensional Prediction and Filtering Problem and the Factorization of Spectral Matrices", J. Franklin Inst., vol. 272, no. 2, pp. 87-99, August 1961.

(482) Wong, E. and M. Zakai, "On Convergence of the Solutions of Differential Equations Involving Brownian Motion", Univ. of California Electron. Res. Lab. Rept. 65-5, Berkeley, Calif., January 1965.

(483) Wong, E. and M. Zakai, "On the Relation between Ordinary and Stochastic Differential Equations", Intern. J. Eng. Sci., vol. 3, pp. 213-229, 1965.

(484) Wong, E., Stochastic Processes in Information and Dynamical Systems, McGraw-Hill Book Co., Inc., New York, 1971.

(485) Wonham, W.M., "Some Applications of Stochastic Differential Equations to Optimal Nonlinear Filtering", J. SIAM Control, ser. A, vol. 2, no. 3, pp. 347-369, 1965.

(486) Woodward, P. M., Probability and Information Theory with Applications to Radar, McGraw-Hill Book Company, New York, 1957.

(487) Wozencraft, J.M. and I.M. Jacobs, Priciples of Communication Engineering, John Wiley & Sons, Inc., New York, 1965.

(488) Yaglom, A.M., Introduction to the Theory of Stationary Random Functions, Englewood Cliffs, N.J., Prentice-Hall, Inc., 1962.

(489) Youla, D.C., "On the Factorization of Rational Matrices", IEEE Trans. Inform. Theory, vol. IT-7, no. 3, pp. 172-189, July 1961.

(490) Young, D.M., Iterative Solution of Large Linear Systems, Academic Press, New York, 1971.

(491) Zadeh, L.A., and C.A. Desoer, Linear System Theory: The State Space Approach, McGraw-Hill Book Company, New York, 1963.

(492) Zadeh, L.A. and J.R. Ragazzini, "An Extension of Wiener's Theory of Prediction", J. Appl. Phys. Applied Physics, vol. 21, pp. 645-655, July 1950.

(493) Zadeh, L.A. and J.R. Ragazzini, "Optimum Filters for the Detection of Signals in Noise", Proc. IRE, vol. 40, pp. 1223-1231, 1952.

Index